MW00440318

Name: _____

Cell/Text Number: _____

Workforce Quality Matters Too!!

...an n-squared value of 1.1? What does that mean?

Ok, well, we'll get back to you on that...

To be honest with you, I don't really know!

He definitely needs to sign-up for a course or two at laserpronet.com

What we do at LaserPronet

- Screening Exams
- Professional Development Courses
- Professional Growth Plans
- Certifications

Laser Advanced Concepts

Course Notes and Workbook

Copyright © 2022

Sukuta Technologies, LLC

All rights reserved

Table of Contents

Topic   Page
1. Laser Technician Series   4
2. Laser Advanced Concepts with Self-tests   7

1. Laser Technician Series

Laser Technician I Job Description: Level I Technicians perform final tests on lasers and laser systems to ensure that they fully comply with customer specifications. All tests are completely documented for both internal and external use.
Environment: Test Laboratory.
Requirements: Must understand laser and optics principles and, laser beam performance specifications. Experience testing/evaluating laser beams and working with photonics test equipment is necessary. Also, ability to follow prescribed written procedures, directions and strict adherence to best laser lab and manufacturing practices are required.
Minimum Educational Requirements: Candidate must hold a certificate/degree in Laser/Electro-Optics Technology or related discipline

Laser Technician II Job Description: Level II Technicians assemble and troubleshoot common optical problems in laser systems. They perform final tests on the systems to ensure that they fully comply with customer specifications and all tests are completely documented for both internal and external use.
Environment: Manufacturing and Test Lab
Requirements: Must understand Wave and Geometrical Optics, Gaussian Beam Propagation, Nonlinear Optics and how acousto- and

electro-optics modulators, optical components and accessories work. Experience aligning laser systems, troubleshooting laser beam aberrations and working with photonics test equipment is a must. Also, ability to follow prescribed written procedures, directions and strict adherence to best laser lab and manufacturing practices are required.

Minimum Educational Requirements: Candidate must hold a certificate/degree in Laser/Electro-Optics Technology or related discipline.

Laser Technician III Job Description: Level III Technicians assemble, align, burn-in, test, and tune/troubleshoot laser heads until they meet all performance specifications.
Environment: Manufacturing
Requirements: Must understand the fundamentals of solid-state laser technology, accessories and support systems. Experience aligning laser systems and cavities/resonators and using photonics test equipment is a must. Also, ability to follow prescribed written procedures, directions and strict adherence to best laser lab and manufacturing practices are required.
Minimum Educational Requirements: Candidate must hold a certificate/degree in Laser/Electro-Optics Technology or related discipline

Technician IV Job Description: Level IV Technicians support Research and Development (R&D) and, customers. Customer Support/Technical Service Technicians work on deployed lasers and laser systems in support of existing customers. R & D/Engineering Technicians support scientists and engineers improve existing and create the next generation laser technologies. Level IV Technicians work under limited supervision.

Environment: Research/Engineering Lab and Field

Requirements: Must have a thorough understanding of solid-state laser technologies and support systems, experimentation and research protocols. Experience collecting and analyzing data, troubleshooting/problem-solving, using MS Office to generate test reports and writing technical reports is required. Customer Support/Service technicians may have to travel to customer sites. **Minimum Educational Requirements:** Candidate must hold a certificate/degree in Laser/Electro-Optics Technology or related discipline.

2. Laser Advanced Concepts with Self-Tests

Topic	Page
2.1 Overview of Nonlinear Optics and Optical Parametric Generation	8
2.2 Overview of Tunable Ultrafast Laser Systems	40
2.3 Overview of ABCD Matrix Optics	63
2.4 Laser Gaussian Beam Propagation and Diffraction-limited Focusing	86

2.1. Overview of Nonlinear Optics and Optical Parametric Generation

2.1.1. Nonlinear Optics and Harmonic Beam Generation
2.1.2. Phase Matching in Nonlinear Crystals
2.1.3. Sum and Difference Frequency Generation
2.1.4. Optical Parametric Generation
2.1.5. Optical Parametric Oscillators
2.1.6. Optical Parametric Amplifiers
2.1.7. Key Terms and Defintions
2.1.8. Self-Test 2.1.

2.1.1. Nonlinear Optics and Harmonic Beam Generation

i. In general, we deal with linear optics everyday i.e. if you input light of a specific wavelength onto a component, e.g. lens, it outputs the same wavelength.

ii. Nonlinear optics is when you input light of a specific wavelength through a component and it outputs a different wavelength from the input wavelength.

- However, dispersion effects may prevent this from happening

under ordinary conditions.

iii. If the fundamantal and harmonic light beams are not phase matched the production of harmonic beams can be challenging or inefficient.

2.1.2. **Phase Matching in Nonlinear Crystals**

i.   Nonlinear crystals are used to extend the wavelength regimes of lasers/laser systems through
    a.   Beam harmonics generation using high-power lasers, and
    b.   parametric oscillators

ii. The power/energy of the harmonic beam, e.g. second harmonic, generated by non-linear crystals is generally very small due to dispersion effects.
    a.   To increase the energy/power output, the two beams involved must share a common index of refraction
    b.   Phase matching is the creation of conditions through which the two beams involved can share a common index of the refraction/phase velocities.

iii. The difference in propagation speed between the inpout beam, e.g. fundamental beam, and the resulting hramonic beam, e.g. second harmonic, in a non-linear crystal means that they will interfere destructively

iv. The solution would be to select a nonlinear crystal whose birefringence exactly compensates for its dispersion resulting in the same index of refraction, n, for e.g. both the fundamental and the SHG beam.

    a.   Specifically and for example for second harmonic generation

$$n_{fundamental} = n_{SHG}$$

- <u>Birefringence</u> – the property of certain materials to have two refractive indices

- <u>Dispersion</u> is the dependence of the index of refraction on wavelength $n = c/v$

v. **Type I Phase Matching** is when the incoming beam has a single polarization and the produced harmonic beam has an **opposite polarization**.

vi. **Type II Phase Matching** is when the incoming beam a has **two** polarization states and the produced **harmonic can have either polarization**

Birefringnet Crystal Phase Matching

a. Phase matching can be achieved using a a bire-fringent crystal that is "tuned" to allow, e.g. the input beam to share a common index of refraction with its harmonic beam.

b. Birefrinenget materials exhibit different indicies of refraction for different polarizations i.e. ordinary (o) and extra-ordinary (e) index of refraction

c. The aforementioned qualities are used to achieve and optimize phase matching.

$$\text{SHG Conversion Efficiency} = \frac{SHG\ Power}{Fundamental\ Power}$$

Conversion Efficiency depends on

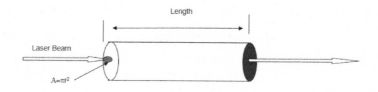

Figure 2.1.1. Illustration of harmonic beam generation set-up and some of the key variables

a. L, length of nonlinear crystal (l^2)

b. P_f, the fundamental power (P_f^2)

    i. Convesrsion effieienceis will therefore be highest/higher

11

        ii. for pulsed compared to cw output.
- c. A , Cross-sectional area of the beam in the non-linear crystal
- d. Phase matching

- As one achieves tighter focusing the contact area decreases and the irradiance on the crystal increases.
- Issues with Fundamental Beam Focusing
  - Continued focusing past the optimum spot size could
    - fracture the harmonic crystal
    - make the fundamental beam highly divergent thus lower the conversion efficiency.
      - However, the incoming fundamental beam will no longer have an ideal plane wave front needed for optimal conversion efficiency.
        - Highest conversion efficiencies are achieved when the beam is collimated/has a planar wavefront.

vii. When perfect phase matching is achieved the transfer of input wave energy is maximized

    a. Phase matching can be achieved by either

        [1] varying the crytsal angle i.e. angle-tuning e.g. KTP or
        [2] its temperature i.e. temp-tuning e.g. LBO, BBO crystals etc..

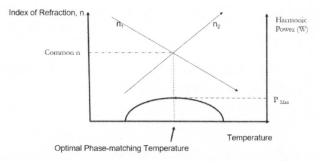

Figure 2.1.2. Illustration of he condition necessary to maximum the conversion efficienecy of harmonic beam generation.

viii. ***In summary, select a birefrinent crystal with the correct nonlinear properties then phase-match until conversion efficiency is optimzed.***

[1] Phase-Matching through Angle-Tuning

a. The crystal can be rotated with respect to the incoming beam until the proper birefringence is obtained.

b. In principle, by adjusting/aligning the crystal orientation we can only achieve perfect phase matching only if the incoming wave has plane waves i.e. collimated beam.

    i. However, real waves do not have plane waves but spherical wave fronts i.e. divergent beams.

c. It is therefore difficult/impossible to achieve perfect phase matching through alignments if the beam is not 100% collimated

13

[2] Phase-Matching through Temp-Tuning
Automated Temperature Tuning Example

- The nonlinear crystals are atumomatically heated/cooled to temperatures where the birefringence exactly compensates for the dispersion.

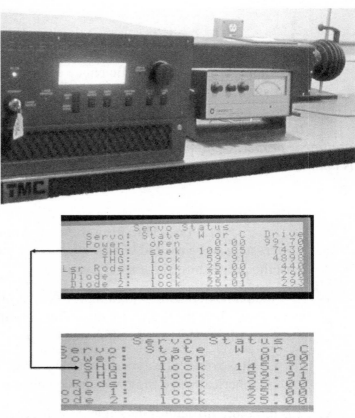

Figure 2.1.3. A laser self temp-tuning to maximize conversion efficiency.

- The SHG crystal is housed in an environment that is at a temperature that ensures that $n_{SHG}=n_{fundamental}$

Personal Notes

2.1.3. Sum and Difference Frequency Generation

- Two input waves of frequency f_1 and f_2 create a third frequency f_3 through a nonlinear interaction

	Input=	Output
SFG	$E_1+E_2 =$	E_3
DFG	$E_1-E_2=$	E_3

Example of Sum Frequency Generation (SFG)

Common Crystals
- Lithium Triborate (LBO),
- Potassium Titanyl Phosphate (KTP),
- Beta-Barium Borate (BBO)

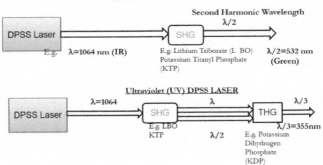

Figure 2.1.4. Illustration of Second and Third Harmonic Generation.

- Second Harmonic Generation (SHG)
  - Also known as frequency doubling
- Third Harmonic Generation (THG)
  - Also known as frequency tripling
- Fourth Harmonic Generation (FHG)

SHG Generation

- **$E=hc/\lambda=hf$**   3.3
  - Where h is Plank's constant,
  - c is the speed of light,
  - and f is frequency

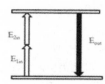

- **$E_{1in}+E_{2in}=E_{out}$**   3.4

- <u>SHG case</u>

$E_{1in}=E_{2in}$

$hc/\lambda_1 + hc/\lambda_1=hc/\lambda_{out}$

Factor out hc $1/\lambda_1 + 1/\lambda_1=1/\lambda_{out}$

$2/\lambda_1=1/\lambda_{out}$

Therefore

$\lambda_{out}=\lambda_1/2$

If $\lambda_1=1064nm$

This therefore means that

$\lambda_{out}=\lambda_1/2=$ 1064nm/2 =**532nm (SHG wavelength)**

Fundamental-SHG Frequency Relationship

$E=hf$ $=hc/\lambda$

$E_{1in}+E_{2in}=E_{out}$

$hf_1+hf_1=hf_{SHG}$

Cancel out h

$f_1+f_1=f_{SHG}$

so

$2f_1=f_{SHG}$

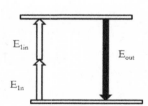

So the SHG (**f_{SHG}**)beam has twice the frequency(**$2f_1$**) , or other words is twice as energetic as the fundamental beam (f_1).

- SHG crystals are sometimes called Frequency Doubling Crystals

17

Frequency Tripling

Starting with

$E_{in} = E_{out}$

$hc / \lambda_2 + hc/\lambda_1 = hc/\lambda_{out}$

"Mix" the following wavelengths to find the output wavelengths

(a) 532nm and 1064nm

$\lambda_{THG} =$

$f_{THG} = \underline{\quad} f_1$

Forth Harmonic Generation - Class Exercise

Starting with

$E_{in} = E_{out}$

$hc / \lambda_2 + hc/\lambda_2 = hc/\lambda_{out}$

Mix the following wavelengths to find the output wavelengths

(a) 532nm and 532nm

$\lambda_{FHG} =$

$f_{FHG} = \underline{\quad} f_1$

Personal Notes

2.1.4. **Optical Parametric Generation**

- One beam λ_{pump} creates two beams λ_{signal} and λ_{idler} through a nonlinear interaction in a resonant cavity/oscillator.

	Input=	Output
Sum	$E_{pump}=$	$E_{signal}+E_{idler}$

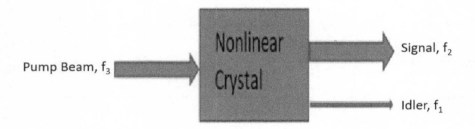

Figure 2.1.5. Optical parametric oscillation and "splitting" of a photon into signal and idler beam beams.

- f=frequency
- Recall the E=hf
- Where E is energy
and f is Plank's constant
- $f_{pump}=f_{idler}+f_{signal}$
- Generally $f_{signal}>f_{iddler}$

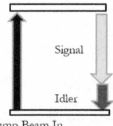

- Generally output λ_{signal} has more power compared to λ_{idler}

2.1.5. Optical Parametric Oscillators

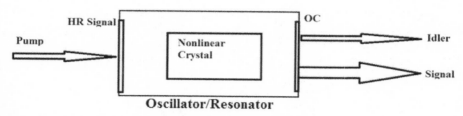

Figure 2.1.6. An optical paramtric oscillator cavity.

- The resonant cavity can be tuned to resonate at either idler or signal wavelenegth or both
  - Phase-matching (angle or temperature tuning) allows for the tuning of both signal and idler beams.
  - The specific wavelengths of the signal and idler depend on the angle at which the pump beam makes with the crystal
    - The highest conversion efficiencies are realized when all the three waves i.e. the pump, idler and signal are phase matched.

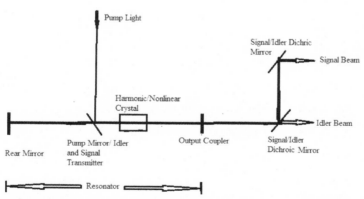

Figure 2.1.7. Schematic layout of an optical paramtric oscillator cavity

21

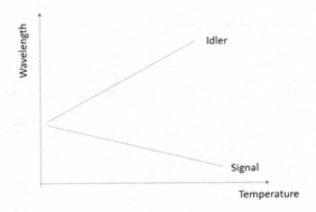

Figure 2.1.8. Phase matching an optical parametric system for output wavelength tunability.

Personal Notes

2.1.6. Optical Parametric Amplifiers

- o If the **signal** is **amplified** by the pump beam the OPO is called an Optical Parametric Amplifier **(OPA)**
- o Recall that the nonlinear crystal is in an oscillator with a HR for the signal f_2

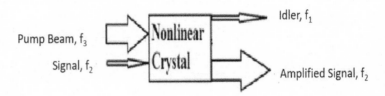

Figure 2.1.9. Illustration of optical parametric amplification.

	Input=	Output
Sum	$E_{signal} + E_{pump} =$	$E_{amplified\ signal} + E_{idler}$

The OPA transfers energy to the signal beam resulting in an amplified signal beam and a weal idler beam,

2.1.7. Key Terms and Defintions

- **BBO,** Beta Barium Borate, is a nonlinear crystal that can be used for laser beam harmonic generation.
- **Conversion Efficiency** is the fraction/percent of the fundamental/first harmonic beam power is being converted to the intended higher harmonic such as second, third, forth etc.. harmonic.
- **Frequency Doubling Crystals** are the same as SHG Crystals.
- **Frequency Tripling Crystals** are the same as THG crystals
- **Fundamental laser beam** is the beam exiting a basic laser cavity slated for harmonics generation(s). Please note that some harmonics generation schemes could be intra-cavity.
- **Harmonic Generation** is when the fundamental, or first harmonic, beam of a laser is converted to its harmonics after passing through a non-linear crystal and/or a series of such crystals.
- **Idler** is the less intense beam after optical parametric generation.
- **KTP**, Potassium Titanium Oxide Phosphate (KTiOPO4) also known as Potassium Titanly Phosphate , is a laser beam harmonics generating crystal.
- **LBO, Lithium Troborate** (LiB$_3$O$_5$) a laser beam harmonics generating crystal.
- **Second Harmonic Generation (SHG)** is when the fundamental wavelength (λ_1) of a laser beam with frequency f_1 is halved ($\lambda_1/2=\lambda_2$) while its frequency is doubled ($2f_1=f_2$).
- **Optical Parametric Amplifier (OPA**) is the process of OPG but whereby one of the two resulting beams/photons (signal or idler) is amplified.Usually the output signal beam is amplified through the introduction of additioanl amplification signal into the oscillator.
- **Optical Parametric Generation (OPG)** is the spliting of a photon/beam into two photons/beams using nonlinear crystals. The photon/beam is split into the idler and signal photons/beams.
- **Optical Paramtric Oscillator (OPO)** is the resonat cavity in which OPG takes place
- **SHG Crystal** is the crystalline material in which second harmonic generation takes place. See Frequency Doubling Crystals.

- **Signal** is the more intense beam resulting from optical parametric generation.
- **THG Crystal** is the crystalline material in which third harmonic generation takes place. See Frequency Tripling Crystals.
- **Third Harmonic Generation (THG)** is when the fundamental wavelength (λ_1) of a laser beam with frequency f_1 is reduced by a third ($\lambda_1/3=\lambda_3$) while its frequency is tripled ($3f_1=f_3$)

Self-Test 2.1. Nonlinear Optics and Optical Parametric Generation.

1. In _____ Optics a wavelength passed through a component is alos output.
   a. Nonlinear
   b. Linear
   c. Any of the above
   d. None of the above

2. In _____ Optics when a wavelength passed through a component a different wavelength is output.
   a. Nonlinear
   b. Linear
   c. Any of the above
   d. None of the above

3. Nonlinear crystals are used to extend the wavelength output of _____ laser beams through harmonic generations and parametric oscillators
   a. low-power
   b. high power
   c. Any of the above
   d. None of the above

4. Conversion Efficiency is _____
   a. $P_{Harmonic}/P_{fundamenal}$
   b. $P_{fundamenal}/P_{Harmonic}$
   c. $P_{Harmonic}/P_{1064}$
   d. A and C
   e. None of the above

5. To increase conversion efficiency of a SHG crystal you can
   a. get a longer crystal
   b. increase fundamental power
   c. decrease the fundamental beam's cross-sectional area
   d. a and b
   e. a, b and c

6. LBO stands for
   a. Lithium Triborate
   b. Potassium Titanyl Phosphate
   c. beta-Barium Borate
   d. Neodymium
   e. None of the above

7. BBO stands for
   a. Lithium Triborate
   b. Potassium Titanyl Phosphate
   c. beta-Barium Borate
   d. Neodymium
   e. None of the above

8. KTP stands for
   a. Lithium Triborate
   b. Potassium Titanyl Phosphate
   c. beta-Barium Borate
   d. Neodymium
   e. None of the above

9. Which one of these is temperature tuned for harmonic beam generation
   a. LBO
   b. KTP
   c. BBO
   d. Nd:YAG
   e. ND:YVO4
   f. a and c
   g. None of the above

10. A THG wavelength of 355 nm can be achieved my mixing the
    _____ beams
    a. 1064 nm
    b. 532 nm
    c. 355nm
    d. a and b
    e. Any of the above
    f. None of the above

11. 4HG (266 nm) of a fundamental 1064 nm laser beam can be achieved my mixing
    a. two second harmonic (532 nm) beams/photons.
    b. two fundamental (1064 nm) beams/photons
    c. two third harmonic (355 nm) beams/photons
    d. any of the above
    e. None of the above

12. A dichroic mirror can be used to _____ beams of different wavelengths fundamental from the second harmonic.
    a. separate
    b. combine
    c. Any of the above
    d. None of the above

13. Phase-matching in a nonlinear crystal is when the _____
    a. input and harmonic beams share a common index of refraction
    b. power output by the harmonic beam is maximized
    c. a and b
    d. None of the above

14. Phase-matching in a nonlinear crystal is generally inhibited by _____ effects.
    a. dispersion
    b. magnetic
    c. chemical
    d. All of the above
    e. None of the above

15. If phase-matched, harmonic light generated in a nonlinaer crystal a one point will interfere _____ with that generated at another point.
    a. destructively
    b. constructively
    c. all of the above
    d. None of the above

16. Type I Phase Matching is when the incoming beam has a single polarization and the produced harmonic beam has a(n)_____ polarization.
    a. Identical
    b. Opposite
    c. any of the above

17. Type II Phase Matching is when the incoming beam a has _____ polarization states and the produced harmonic beam can have either of those polarizations
    a. Two
    b. Three
    c. Four
    d. Any of the above
    e. None of the above

18. Phase matching crystals _____
    a. are birefringemt
    b. can have their indices of refaction " tuned".
    c. have two indices of refraction
    d. a and b
    e. all the above
    f. None of the above

19. Birefrinenget crystals indices of refraction are called _____
    a. ordinary (o)
    b. extra-ordinary (e)
    c. polar (p)
    d. a and b
    e. All the above
    f. None of the above

20. Phase matching can be achieved by
    a. angle-tuning
    b. Temperature-tuning
    c. Mode-matching
    d. a and b
    e. All the above
    f. None of the above

21. When the fundamenatl beam is tightly focused onto a
    nonlinear/harmonic crystal the _____ increases.
    a. contact
    b. irradiance
    c. a and b
    d. all of the above
    e. None of the above

22. Tightly focusing the fundamental/input beam onto a
    nonlinaer/harmonic crystal can lead to _____ .
    a. the crystal fracture
    b. highly divergence of the fundamenatl mode
    c. a and b
    d. None of the above.

31

23. The conversion efficiency of the harmonic generation process can be optimized if the incoming fundamental beam _____.
   a. has planar wavefronts
   b. is collimated
   c. a and b
   d. All the above
   e. None of the above

24. _____ is when a nonlinear/harmonic crytsal crystal is heated/cooled to a temperature where the birefringence exactly compensates for the dispersion.
   a. Temp-tuning
   b. Angle-tuning
   c. a and b
   d. None of the above

25. _____ is when a harmonic/nonlinear crystal rotated with respect to the incoming beam until the proper birefringence is obtained.
   a. Temp-tuning
   b. Angle tuning
   c. a and b
   d. None of the above

26. In a _____ process two input waves of frequency f_1 (or wavelength λ_1), and f_2 (or waveklength) λ_2, create a third frequency f_3 (or wavelemgth λ_3) through a nonlinear interaction.
   a. Sum Frequency Generation (SFG)
   b. Difference Frequency Generation (DFG)
   c. a and b
   d. None of the above

27. A technician's workplan in a manufacturing company inlcude
_____ .
    a. assembling components
    b. aligning laser beams through components
    c. optimizing laser/laser stystems performance
    d. testing performance
    e. all the above
    f. None of the above

28. _____ is/are a nonliner crystal(s) that can be used harmonic generation depending on wavelength.
    a. BBO
    b. KTP
    c. LBO
    d. a and b
    e. All the above
    f. None of the above

29. _____ Efficiency is the fraction/percent of the input beam power is being converted to into a higher harmonic.
    a. Conversion
    b. Wall-plug
    c. Slope
    d. a and b
    e. b and c
    f. All the above
    g. None of the above

30. Frequency Doubling Crystals are the same as _____ crystals.
    a. SHG
    b. THG
    c. FHG
    d. B and C
    e. All the above

31. Frequency Tripling Crystals are the same as _____ crystals.
    a. SHG
    b. THG
    c. FHG
    d. B and C
    e. All the above

32. The fundamental laser beam is the beam that exits the _____ to harmonics generation set-up.
    a. basic laser cavity
    b. Fourth Harmonic Generatror
    c. Q-switch
    d. A and B
    e. All the aboive
    f. None of the above.

33. _____ is when an input beam is converted to higher harmonics after passing through a non-linear crystal and/or a series of such crystals.
    a. Harmonic Generation
    b. Phase-matching
    c. Mode-matching
    d. B and C
    e. A and C
    f. None of the above
    g. All the above

34. _____ is/are a laser beam harmonics generating crystal(s).
    a. KTP
    b. LBO
    c. BBO
    d. THG
    e. a and b
    f. a,b, and c
    g. All the above
    h. None of the above

35. _____ Crystal is the crystalline material in which third harmonic generation takes place.
   a. THG
   b. SHG
   c. FHG
   d. LBO
   e. A and C
   f. All the above
   g. None of the above

36. Optical Parametric Generation (OPG) takes place in a _____ crystal.
   a. nonlinear
   b. active medium
   c. birefrinengt
   d. a and c
   e. All of the abive
   f. None of the above

37. During OPG two beams/photons are _____
   a. combined
   b. formed
   c. a and b
   d. None of the above

38. In OPG the output beam(s) is/are _____ beams(s).
   a. pump
   b. Idler
   c. Signal
   d. a and c
   e. b and c
   f. All of the above
   g. None of the above

39. During OPG _____ energy is equal to the pump energy
   a. signal
   b. idler
   c. third Harmonic
   d. a and b
   e. All of the above
   f. None of the above

40. OPG generally takes place on a(n) _____
   a. oscillator
   b. Q-switch
   c. a or b
   d. All of the abive
   e. None of the above

41. OPO stands for _____
   a. Optical Parametric Optics
   b. Optical Parametric Oscillator
   c. Optical Parametric Oscillations
   d. b and c
   e. All the above
   f. None of the above

42. An OPO can resonant the _____ signal(s)
   a. idler
   b. signal
   c. pump
   d. a and b
   e. All of the above
   f. None iof the above

43. Phase-matching (angle or temperature tuning) allows for the tuning
    of _____ beam(s)
    a. signal
    b. idler
    c. pump
    d. a and b
    e. all the above
    f. None of the above

44. The highest conversion efficiencies in an OPO are possible when
    _____ beams(s) are phase matched
    a. pump,
    b. idler
    c. signal
    d. b and c
    e. All the above
    f. None of the above

45. In an Optical Parametric Amplifier the _____ beam is/are
    generally amplified
    a. pump
    b. signal
    c. idler
    d. b and c
    e. All the above
    f. None of the above

The next two questions are based the laser head characterized below.
You are to assemble a laser head from the clean room into a laser system. As you receive the laser you note, among other things, that it has an M^2 value of 1.3.

46. As you align the laser system you find out that it is outputting the second and third harmonics of 1064 nm but the system only needs the third harmonic. Which dichroic mirror/reflector can you use to get rid of 1064 nm and the second harmonic?

    a. 1064 nm dichroic mirror/reflector
    b. 355 nm dichroic mirror/reflector
    c. 532 nm dichroic mirror/reflector
    d. 266 nm dichroic mirror/reflector

47. If after retaining the third harmonic (from the previous question) you realize that the beam is highly divergent, which optical component/system you would utilize to fight/reduce the divergence.
    a. microscope
    b. diffraction grating
    c. collimator
    d. convex lens
    e. None of the above

48. THG (355 nm) can be achieved my mixing the fundamental (1064 nm) with the _____ harmonic

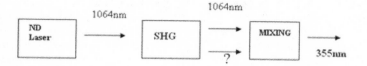

Figure T1.1. Third Harmonic Generation?

    a. Second harmonic
    b. Third harmonic
    c. Forth harmonic
    d. All the above
    e. None of the above

49. 4HG (266 nm) of a 1064 nm laser beam can be achieved my mixing two _____ beams.
    a. Second harmonic
    b. Third harmonic
    c. Forth harmonic
    d. All the above
    e. None of the above

Laser Advanced Concepts

2.2. *Overview of Tunable Ultrafast Laser Systems*

2.2.1. Ti:Sa Broadband Tunable Lasers
2.2.2. Mode-locking Broadband Lasers Systems for Ultra-Fast Pulses
2.2.3. Mode-locking Methods
2.2.4. Mode-locked Output Pulse Sampling
2.2.5. Broadband and Ultrashort Pulse Beams' Propagation Issues
2.2.6. Key Terms and Definitions
2.2.7. Self-Test Tunable Broadband Δf Laser Systems

2.1. Ti:Sa Broadband Tunable Lasers

- Even though there are other solid state crystals that give broadband spectra Titanium Sapphire, **Ti:Sa**, has emerged as the most robust and suitable for commercial applications. We shall therefore Ti:Sa from here on, including the next section on ultrafast mode-locked lasers

- Titanium, Ti, is a transition metal used as an active ion/species in tunable solid-state lasers.

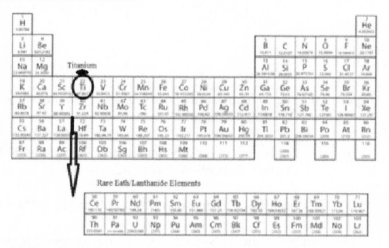

- o Note that Titanium is from a transition metal group
  - Most solid state-laser active media in use are from the rare earth/lanthanide elements e.g. Nd, Ho etc...
  - The lanthanide elements are not tunable hence emit monochromatic laser beams

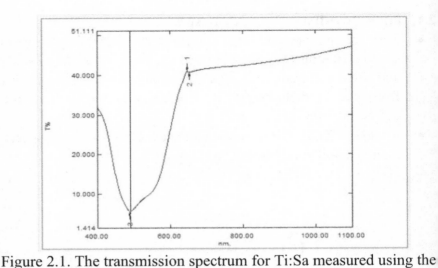

Figure 2.1. The transmission spectrum for Ti:Sa measured using the Shimadzu UV 1800 spectrophotometer.

- o can be pumped over any wavelength between 400 to 600 nm but its peak absorption wavelength is at 488 nm

- o has emission/tuning range from 660 nm to 1090 nm

- o Currently the Ti:Sa lasers are mostly pumped by green lasers usually the green beam is a SHG of a rare-earth solid-state laser such as Nd:YAG, Nd:YVO$_4$ (532 nm) etc.

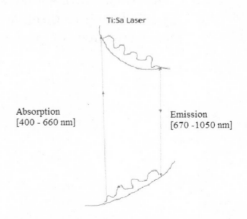

Figure 2.2. Ti:Sa vibronic absorption and emission of electromagnetic energy.

42

- Laser action is 4-level **vibronic** i.e. both vibrational and electronic energy levels are involved in the laser radiation transitions.
- The broadband output range is from 670 to about 1050 nm
  - One wavelength is generally output at a time, the lasers are also called or referred to **Tunable**
- Emission peak @ around 795 nm

- Ti:Sa laser tunable broadband output challenged the "dirty" dye lasers to almost extinction now.

Personal Notes

2.2.2. Mode-locking Broadband Lasers Systems for Ultra-Fast Pulses

- Titanium-doped sapphire crystals ($Ti:Al_2O_3$ or Ti:Sa) has a broad, Δf, output bandwidth it is capable of outputting very shorter, or ultrafast, $\Delta\tau$ pulses such that

$$\Delta\tau\Delta f => .44$$

- Ordinary lasers are free running lasers i.e. the longitudinal modes have random phases and magnitudes
    - Each longitudinal mode oscillates independent of the others with phases spanning from π to $-\pi$ radians in the frequency domain
    - The resulting laser output is time averaged statistical mean value

- On the other hand, mode-locked lasers' longitudinal modes exhibit zero phase difference between each other in frequency space

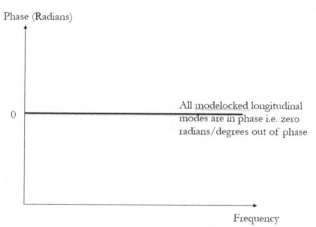

Phase (Radians)

0

All modelocked longitudinal modes are in phase i.e. zero radians/degrees out of phase

Frequency

Figure 2.3. Phase relation in the frequency space in a mode-locked laser.

- The resulting pulsed laser output is a superposition of all the broadband wavelengths exhibiting a Gaussian shape in time space

45

How Mode-locking Modulators Work

- A mode locking modulator works as a fast-optical switch
- Mode-locking is a result of wave beats, as beats in sound waves
- The modulator "locks" the longitudinal modes in the beat wave to simultaneously oscillate in phase

- Mode-locking establishes a fixed-phase relationship among the longitudinal modes in a resonator
    - The broadband longitudinal modes will exhibit collective constructive and destructive interference pattern i.e. the beat wave.
    - Where there is the least destructive and most constructive interference would be the beat wave maximum and vice-versa
        - i. The beat wave maxima oscillate back and forth in the resonator of length L and at speed c
        - ii. The beam wave maximum occurs in a periodic fashion at 2L/c in time space

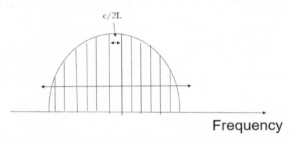

Figure 2.4. Longitudinal modes distribution in the frequency space in a mode-locked laser.

- iii. The only longitudinal modes that can circulate between the mirrors are only those that are mode-locked into a pulse
    1. This results in many longitudinal modes oscillating at a fixed phase
    2. Every time (2L/c) the beat maxima hit the OC there would be a pulse output through the OC

46

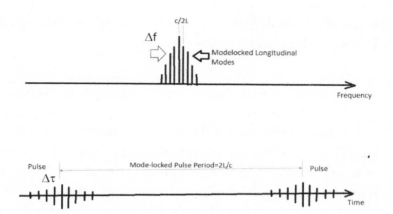

Figure 2.5. Illustration of broadened pulses in spectral and temporal space.

- o In the frequency domain, the mode-locked pulses
  - have discrete spectral intensities/lines
  - have a Gaussian distribution
  - are spaced at c/2L intervals/PRR where L is the length of the cavity and c is the speed of light
- o In the time domain, the output is a multispectral but phase-locked single Gaussian pulse

- The gate/modulator opens once per round-trip transit time, letting the pulse through and closed for the rest of the time.
- The resulting output pulse, i.e. beat max, is an intense multispectral pulse output

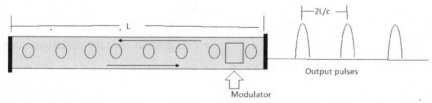

Figure 2.6. Illustration of mode-locked multispectral beats as laser output pulses output.

- In a mode-locked laser the product of Pulse Duration, $\Delta\tau$, and Frequency Bandwidth, Δf, is invariant

$$\Delta\tau\Delta f \Rightarrow .44$$

  - .44 is called the pulse width-bandwidth or time-bandwidth product of a pulse
- This equation tells us that the larger the output bandwidth, **Δf,** of an amplifying medium the shorter, **$\Delta\tau$,** are pulses it will produce
- Since Ti:Sa has a huge bandwidth **Δf**, its mode-locked pulse is typically in the femto-second (10^{-15}s) regime **$\Delta\tau$**.
  - Q-switch pulses of rare earth lasers, e.g. Nd:YAG, typically are in the nanosecond regime (10^{-9}s)

2.2.3. Mode-locking Methods

- **Active Modulators**
  - Active mode-locking requires insertion of modulators inside the resonator
    - Electro-Optical (E-O)
    - Acousto-Optical (A-O)
  - The modulators are activated through external signals.
- **Passive Mode-locking** techniques include SESAM and Kerr-len mode locking. In general, they self-modulate as they naturally react to the laser "light".
  - Semiconductor Saturable Absorber Mirrors **SESAM** SESAM are used as HR mirrors for mode-locking. The SESAM HR cyclically changes reflectivity as it reacts to the laser light coming from the active medium. It basically changes from unsaturated to saturated absorber as a reaction to the laser beam intensity resulting in mod locked pulses

Figure 2.7. Illustration o SESAM passive mode-locking.

- Kerr-Lens Mode-locking. It is a nonlinear phenomenon that exploits self-focusing of high intensity beams as they pass through a laser active medium.
- Dye Cells

Personal Notes

2.2.4. Mode-locked Output Pulses Sampling

- Mode-locked pulses are typically in the pico- femto-second regime
  - For example, a tera Hertz (10^{12}) bandwidth translates to a pico second (10^{-12}) pulse laser.
- Optical, not electronic, techniques are required for measuring pulse durations shorter than picoseconds
  - Fast electronics cannot measure pulse durations shorter than picoseconds.
- Autocorrelators, not oscilloscopes (electronics), are used to measure the temporal characteristics of femto-second pulses
  - Note that Q-switching produces pulses in the nanosecond regime and oscilloscopes are generally used.

2.2.5. Broadband and Ultrashort Pulse Beams' Propagation Issues

B5.1 Pulse Broadening Due Dispersion/Chromatic Aberration

B5.2 Laser Amplifier Damage Control

2.2.5.1 Pulse Broadening Due Dispersion/Chromatic Aberration

- Broadband spectra output by Ti:Sa, and similar lasers, are highly susceptible to dispersion as they propagate through different intra cavity components/materials
- Long wavelengths (e.g. in a visible beam, red) travel faster than short wavelengths (e.g. in a visible beam, blue)

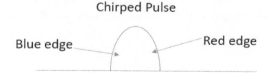

Figure 2.8. Illustration of a chirped pulse resulting from spectral/longitudinal dispersion.

51

- Pulses would be chromatic aberration broadened with longer wavelengths leading shorter wavelengths in the pulse.
- Dispersion/chromatic aberration correction optics are needed to so that the shorter wavelengths can catch-up with longer wavelengths.

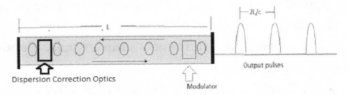

Figure 2.9. Inclusion of dispersion correction optics in a broadband laser

- Optics such as prisms are used to correct/compensate for dispersion/chromatic aberration effects thus allow the short-wave components to catch up with the long wave components.

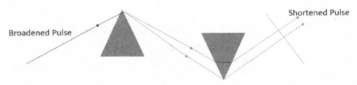

Figure 2.10. Illustration of how dispersion prisms slow down faster wavelengths so that slow ones can catch-up.

- Chirped mirrors
- Diffraction Gratings

2.2.5.2 Laser Amplifier Damage Control
- If amplification is needed pulses are stretched first, amplified while stretched, and then compressed back to original or desired length

2.2.6. Key Terms and Concepts

Anamorphic optics are optics that only affect one, out of 2 (x-y), axis of beam propagating beam in the z-direction.

Bandwidth is range/amount of wavelengths/frequencies.

Broadband Laser Output, is laser output with a large bandwidth, see Bandwidth

Chirped laser pulse, a broadband/polychromatic pulse whose wavelengths have been separated due to dispersion effects where long wavelengths, e.g. red, lead while short ones, e.g. blue, trail as they propagate through optical components.

 Dispersion, the dependence of light wavelengths propagation on index of refraction such that long wavelengths lead while short wavelengths trail. Ordinary/daily observation of dispersion phenomena include the rainbow.

Mode-locking is the "locking" of polychromatic longitudinal waves, such as those originating from a broadband laser active media such as Ti:Sa, in a resonator to oscillate in phase.

  Active mode-locking is when an electronic signal is sent to the modulator to affect the mode-locking, for example A-O and E-O Modulators.

  Passive mode-locking is when a material, such as SESAM, effects mode locking naturally.   ,

SESAM, a passive mode-locking mirror (HR)

Stretched Laser Pulse, a pulse whose peak is lowered while its pulse duration is extended. Generally, this is done while a pulse is being amplified to avoid optical damage in the amplifier. The pulse is compressed as soon as it leaves the amplifier hence its peak goes up while its pulse duration is lowered.

Ti:Sa, a common and almost commercially exclusive broadband laser active medium. Its broadband output makes it suitable for the generation of ultrashort pulses through mode-locking.

Tunable Laser Output, the ability to select specific wavelengths for output off available broadband laser output wavelengths. Generally, only one wavelength is output at a time.

Ultrafast Laser Pulse, laser pulses must be faster/shorter than nanosecond pulse duration, the norm or reference, and are typically in the femtosecond regime. Ultrafast output generally occurs in mode locked Ti:Sa lasers.

Vibronic laser transitions, laser output that involve both molecular vibrational and atomic electronic energy transitions such as in Ti:Sa

2.2.7. Self-Test Tunable Broadband Δf Laser Systems

1. _____ has emerged as the most suitable active-medium for broadband tunable lasers
   a. Nd:YAG
   b. ND:YLF
   c. b and c
   d. All the above
   e. None of the above

2. _____ is/are a rare earth active medium
   a. Ti:Sa
   b. Nd:YAG
   c. Nd:YLF
   d. b and c
   e. All the above
   f. None of the above

3. _____ is/are a transition metal active medium
   a. Ti:Sa
   b. Nd:YAG
   c. ND:YLF
   d. b and c
   e. All the above
   f. None of the above

4. _____ exhibits vibronic electronic transition
   a. Ti:Sa
   b. Nd:YAG
   c. Nd:YLF
   d. b and c
   e. All the above
   f. None of the above

5. An active medium with a broadband output can also have_____ output
   a. ultrafast
   b. tunable
   c. a and b
   d. None of the above

6. _____ lasers output ranges from about 700 to about 1100 nm
   a. Ti:Sa
   b. Nd:YAG
   c. ND:YLF
   d. b and c
   e. All the above
   f. None of the above

7. Bandwidth is _____.
   a. The range of wavelengths or frequencies over which a laser operates
   b. The lowest energy level of an atom or molecule
   c. The quantized amounts of energy that can be stored in an atom or molecule
   d. All the above
   e. None of the above

8. Angstrom is _____.
   a. 10^{-9}
   b. 10^{-15}
   c. 10^{-10}
   d. All the above
   e. None of the above

9. Micron is _____.
   a. 10^{-15}
   b. 10^{-6}
   c. 10^{-15}
   d. All the above
   e. None of the above

10. Nano is _____.
    a. 10^{-9}
    b. 10^{-12}
    c. 10^{-10}
    d. All the above
    e. None of the above

11. Pico is _____.
    a. 10^{-9}
    b. 10^{-12}
    c. 10^{-15}
    d. All the above
    e. None of the above

12. femto is ---------.
    a. 10^{-9}
    b. 10^{-12}
    c. 10^{-15}
    d. All the above
    e. None of the above

13. A 1064 nm laser beam has a wavelength of
    _____ μm.
    a. 1.064
    b. .1064
    c. 10.64
    d. Any of the above
    e. None of the above

14. In a _____ longitudinal modes have random phases and magnitudes
    a. free running
    b. mode-locked
    c. All the above
    d. None of the above

15. _____ longitudinal modes exhibit zero phase difference between each other in frequency space
    a. Free running
    b. Mode-locked
    c. All the above
    d. None of the above

16. Mode-locked longitudinal modes_____ form
    a. beat waves
    b. oscillate in phase
    c. All the above
    d. None of the above

17. A mode locked laser pulse period is given by _____.
    a. 2L/c
    b. c/2L
    c. a or b
    d. All the above
    e. None of the above

18. A mode-locked laser pulse generated from a broadband active medium is _____.
    a. multispectral
    b. monochromatic
    c. polychromatic
    d. a and c
    e. All the above
    f. None of the above

19. Active modulators _____ require external controls to modulate a beam
    a. do
    b. do not
    c. Depends on the specific modulator

20. Passive modulators _____ require external controls to modulate a beam
    a. Do
    b. do not
    c. Depends on the specific modulator

21. Active modulators include _
    a. Electro-Optical (E-O)
    b. Acousto-Optical (A-O)
    c. SESAM
    d. Kerr Lens
    e. a and b
    f. All the above
    g. None of the above

22. Passive modulators include _____.
    a. Electro-Optical (E-O)
    b. Acousto-Optical (A-O)
    c. SESAM
    d. Kerr Lens
    e. c and d
    f. All the above
    g. None of the above

23. Ti:Sa mode-locked laser pulses are typically in the _____ second regime
    a. milli-
    b. nano
    c. femto
    d. pico
    e. c and d
    f. all the above
    g. None of the above

24. A tera Hertz (10^{12}) bandwidth translates to a _____ second laser pulse.
    a. nano
    b. pico
    c. femto
    d. b and c
    e. all the above
    f. None of the above

25. _____ are used to measure the temporal characteristics of femto-second pulses
    a. Oscilloscopes
    b. Power meters
    c. Autocorrelators
    d. a and c
    e. All the above
    f. None of the above

26. Broadband spectra or pulse are highly susceptible to _____ as they propagate through different optical components/materials.
    a. Polarization
    b. chromatic aberration/dispersion
    c. extinction
    d. b and c
    e. all the above
    f. None of the above

27. Broadband mode-locked pulses are
_____ by chromatic aberration
   a. broadened
   b. shortened
   c. unaffected
   d. None of the above
   e. All the above

28. In a broadened pulse spectrum _____.
wavelengths lead
   a. long
   b. short
   c. average
   d. Any of the above
   e. None of the above

29. _____ optics are used to correct for
pulse broadening.
   a. Anamorphic
   b. Hardened
   c. Polarization
   d. All the above
   e. None of the above

30. Anamorphic optics in ultrafast/broadband
lasers include a _____.
   a. one dispersion prism
   b. dispersion prism pair
   c. b and c
   d. all the above
   e. None of the above

31. Insertion of anamorphic optics inside ultrafast/broadband lasers allows for short-wave components to _____ so that they can catch-up with the long wave components.
    a. slow-down
    b. speed-up
    c. stop
    d. Any of the above
    e. None of the above

32. A spectrally broadened pulse is called a _____ pulse
    a. chirped
    b. clumped
    c. smeared
    d. all the above
    e. None of the above

33. Optical damage can be avoided in a laser cavity if a pulse is _____ upon exit.
    a. first stretched, amplified, then compressed
    b. first amplified, stretched, then compressed
    c. first compressed, stretched, then amplified
    d. any of the above
    e. None of the above

34. _____ lasers output ranging from 700 to about 1100 nm
    a. Ti:Sa
    b. Nd:YAG
    c. ND:YLF
    d. b and c
    e. All the above
    f. None of the above

62

Laser Advanced Concepts

2.3. Overview of ABCD Matrix Optics

Topics
2.3.1. Overview of ABCD Matrix Optics Characterization Matrices
2.3.2. The Three Focal Lengths of a Lens
2.3.3. Application Examples of ABCD Matrices to Laser Resonators
2.3.4. Self-Test of ABCD Matrix Optics

2.3.1. Overview of ABCD Matrix Optics Characterization Matrices

We shall assume ideal or paraxial optics thus we will ignore
- aberrations
- assume small angles of incidence such that $\sin\theta = \theta$

Light rays can be traced forward using a 2x2 unimodular matrix

$$\begin{bmatrix} y_2 \\ \Theta_2 \end{bmatrix} = \begin{bmatrix} A & B \\ C & D \end{bmatrix} \begin{bmatrix} y_1 \\ \Theta_1 \end{bmatrix}$$

The column vector represents the object

$$\begin{bmatrix} y_1 \\ \Theta_1 \end{bmatrix}$$

where y_1 is the height and Θ_1 is the angle at which the ray is projected.

This 2x2 square matrix

$$\begin{bmatrix} A & B \\ C & D \end{bmatrix}$$

is a matrix representation of the optical system traversed by the rays.

The column vector represents the image

$$\begin{bmatrix} y_2 \\ \Theta_2 \end{bmatrix}$$

where y_2 is the height and Θ_2 is the angle at which the ray strikes the image plane.

Rays can also be traced backwards through the inversion of the forward matrix as shown below.

$$
\begin{bmatrix} y_1 \\ \Theta_1 \end{bmatrix} = \begin{bmatrix} A & -B \\ -C & D \end{bmatrix} \begin{bmatrix} y_2 \\ \Theta_2 \end{bmatrix}
$$

Common Ray Tracing Matrices
1. Transfer/Translational Matrix, T
2. Refraction Matrix, R
3. Lens Matrix, L
4. Mirror Matrix, M

1. Transfer Matrix, $\boldsymbol{T} = \begin{bmatrix} 1 & T \\ 0 & 1 \end{bmatrix}$

$T = d/n$

Where d is the length of gap through which the rays were transferred and n is the index of refraction of the gap.

2. Refraction Matrix, $R = \begin{bmatrix} 1 & 0 \\ -(n_2-n_1)/r & 1 \end{bmatrix}$

Where n_1 and n_2 is the indices of refraction in the incidence and refracting media respectively, and r is the radius of curvature of the refracting medium.

An equivalent representation of the refraction matrix is shown below.

$$
R = \begin{bmatrix} 1 & 0 \\ -(n_2-n_1)/r\, n_2 & n_1/n_2 \end{bmatrix}
$$

3. Lens Matrix, $L = \begin{bmatrix} 1 & 0 \\ -1/f & 1 \end{bmatrix}$

65

Where f is the focal length of the lens and the power of lens, P, is given by

$$P = \frac{1}{f}$$

4. Mirror Matrix, M=
$$\begin{bmatrix} 1 & 0 \\ 2n/r & 1 \end{bmatrix}$$

Where n is the index of refraction and r the radius of curvature of the mirror

Ray Tracing
Given an optical system in which a ray propagates from left to right from E_1 through E_N then system matrix, S, is given as

$$S = E_N \times E_{N-1} \times E_{N-2} \ldots \ldots E_1$$

for example, given an optical system shown below then the system matrix would be

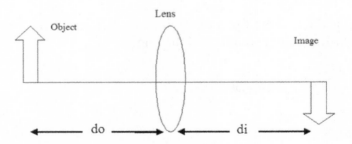

Figure 3.1. An optical imaging system

$$S = \begin{bmatrix} 1 & di \\ 0 & 1 \end{bmatrix} \begin{bmatrix} A & B \\ C & D \end{bmatrix} \begin{bmatrix} 1 & do \\ 0 & 1 \end{bmatrix}$$

Optical System Matrix Properties

We shall now look at the physical significance of the four elements of the system A, B, C, and D

$$
\begin{bmatrix} y_2 \\ \Theta_2 \end{bmatrix} = \begin{bmatrix} A & B \\ C & D \end{bmatrix} \begin{bmatrix} y_1 \\ \Theta_1 \end{bmatrix}
$$

It can be shown that
$y_2 = Ay_1 + B\Theta_1$

If A=0
then
$y_2 = B\Theta_1$

Θ_1 input collimated rays

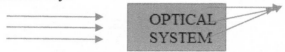

All output rays cross at a height y_2 in the second focal plane

$y_2 = Ay_1 + B\Theta_1$
If B=0
$y_2 = Ay_1$

All input rays originate from the same height y_1

All output rays cross at a height y_2

This is an object-image system and it is also apparent that the magnification of the system is $A = y_2/y_1$

It can be shown that $\Theta_2 = Cy_1 + D\Theta_1$

If C=0

collimated outputs rays
$\Theta_2 = D\Theta_1$

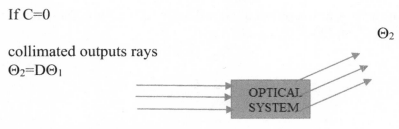

Θ_1 collimated input rays.

The optical system is afocal since it has no focusing power. We can think of D ($=\Theta_2/\Theta_1$) as the angular magnification of the system.

If D=0, then $\Theta_2 = Cy_1$

outputs rays

Θ_2 collimated

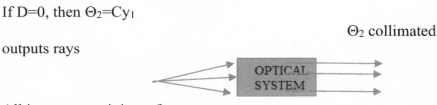

All input rays originate from
the same height y_1 in the first focal plane

The Cardinal Points of an Optical System

1. The Two Focal Points
2. The Two Nodal Points
3. The Two Principal Planes

1. The Two Focal Points

The two focal points were identified in the proceeding section where we showed that

a. The first focal point is located where diverging rays traversing the system exit as parallel rays (when D=0). We shall call this point F_1.
b. The second focal point is located where collimated/parallel rays traversing the optical system converge to upon (when A=0). We shall call this point F_2.

2. The Two Nodal Points

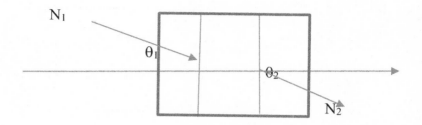

The nodal points are such that any rays directed at first nodal point, N_1, appear as a ray coming from second nodal point, N_2, and both rays make the same angle with the optical axis ($\theta_1 = \theta_2$).

3. The Two Principal Planes

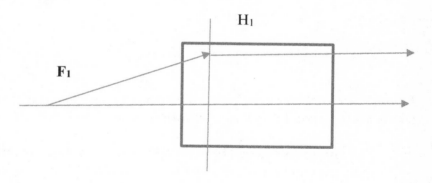

The first principal plane, H_1, is a mathematical point at which diverging rays from the first focal point (F_1) become collimated. In real optical systems, light rays gradually bend in a manner consistent with Snell's Law.

The second principal plane, H_2, is mathematical point at which collimated rays from infinity bend toward the second focal point (F_2).

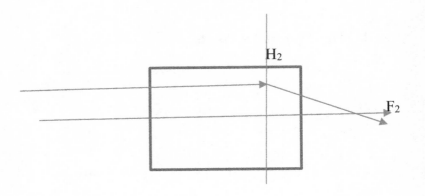

The Effective Focal Lengths of a Lens.

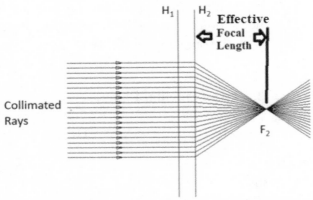

Figure 3.2. Computer simulation of a positive lens using ABCD matrices showing its H_2 and F_2. *Graphics produced using ASAP™ (Breault Research Organization).*

The principal plane, H_1, of an optical system, e.g. lens, may be located within it which can make it prohibitive to measure the focal length.

The back focal length is the distance from F_1 to the vertex/edge of the lens, while the front focal length is the distance from the back edge of the lens to F_2. The focal lengths are measured from the principal planes, H_1 and H_2, to the focal points, F_1 and F_2, is usually referred to as the effective focal length.

Locating the Cardinal Points of a System

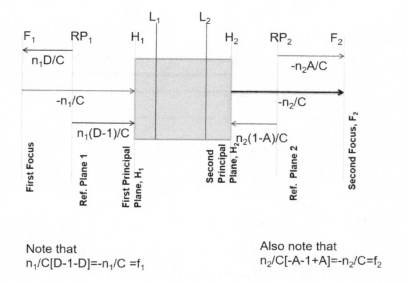

Note that
$n_1/C[D-1-D]=-n_1/C =f_1$

Also note that
$n_2/C[-A-1+A]=-n_2/C=f_2$

Figure 3.4. Encapsulation of ABCD matrix characterization of optical system.

3.2. The Three Focal Lengths of a Lens

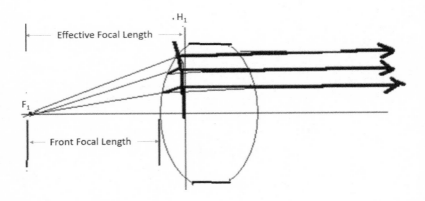

Figure 3.5. Depiction of a positive lens showing the effective and back focal lengths.

    a. Effective Focal Length (EFL)
    b. Front Focal Length (FFL)
    c. Back Focal Length (BFL)

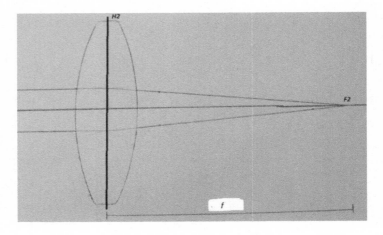

Figure 3.6. The location of H is not discernible within a lens. Generally, the center of a lens is assumed to be the be the location of H, which makes most experimental findings merely approximations of the Effective Focal Length (EFL) of a lens. ABCD Matrix Methods can be used accurately determine the effective focal length pf a lens.

72

3.2.1. Laboratory Exercise: Determination of the Matrix Elements of an Imaging System

Objective: The objective is to experimentally determine of the matrix elements of an optical system

Theory:

The matrix below characterizes our optical system

$$\begin{bmatrix} y_2 \\ v_2 \end{bmatrix} = \begin{bmatrix} A & B \\ C & D \end{bmatrix} \begin{bmatrix} y_1 \\ v_1 \end{bmatrix}$$

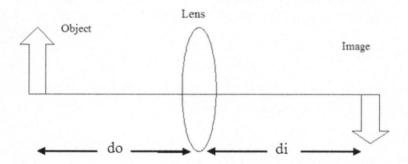

Figure 3.6. Experimental set-up for determining the matrix elements (A, B, C and D) of an object-image system.

d_o is the distance from Object to lens
d_i is the distance from image to lens
OH is the Object height
IH is the image height

73

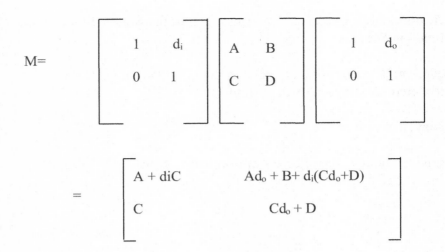

$$M = \begin{bmatrix} 1 & d_i \\ 0 & 1 \end{bmatrix} \begin{bmatrix} A & B \\ C & D \end{bmatrix} \begin{bmatrix} 1 & d_o \\ 0 & 1 \end{bmatrix}$$

$$= \begin{bmatrix} A + d_iC & Ad_o + B + d_i(Cd_o + D) \\ C & Cd_o + D \end{bmatrix}$$

We know that the matrix determinant is unity, the upper right-hand element $(Ad_o + B + d_i(Cd_o + D))$ is equal to zero due the image-object relationship. The upper left-hand element $(A + d_iC)$ is the transverse magnification and the lower right-hand element $(Cd_o + D)$ is the reciprocal and we shall denote it by α

Procedure:
Measure object and image distances do and d_i ten times
Measure object and image heights ten times

Computations:

Finding Elements C and D

Compute the ratio object height/image heights. (Note that if image is inverted then image height should be negative.
Object height/image height$=\alpha= Cd_o + D$
Plot a graph of α against do for the ten data points and find the slope.

Analyze the graph as if it were that of $y = mx + b$ (equation of straight line).

In this case, the slope (m) of the graph will give us matrix element C and the y-intercept would be equivalent to the matrix element D. This will therefore allow us to determine the optical systems elements C and D.

Finding Elements, A and B

Since the upper right-hand element vanishes, we have $Ad_o+B=-d_i(Cd_o+D)$

But $Cd_o +D=\alpha$ hence

$Ad_o + B=-d_i\alpha=\beta$. Since we know d_i and α this means that β can be determined. Plot a graph of β against do for the ten data points and find the slope. Analyze the graph as if it were that of $y=mx + b$ (equation of straight line). In this case, the slope (m) of the graph will give us matrix element A and the y-intercept would be equivalent to the matrix element B. This will therefore allow us to determine the optical systems elements A and B

Data and Computations

	d_o	d_i	OH	IH	$\alpha =OH/IH$ $= Cd_o+D$	$\beta=-d_i*\alpha$
1						
2						
3						
4						
5						
6						
7						
8						
9						
10						

Determinant Test

Find the determinant (AD-BC) and it must be approximately equal to 1. You shall compare your answer (ans) to the expected value

Calculate % Error

$\% Error=|1-ans|/1 \times 100$

Personal Notes

2.3.3. Application Examples of ABCD Matrices to Laser Resonators

2.3.3.1. Matrix Representation of a Laser Resonator

Below we show a typical arrangement in a laser resonator

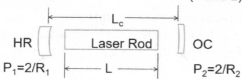

RP(1 and 2)

- The laser rod can be treated as a plane-parallel plate so the T-matrix would have a reduced thickness
- $T=(L_c-L)/1+L/n=L_c-(n-1)L/n$ (Verify this)

Lets make Reference Plane 1 , RP_1, coincide with and Reference Plane 2, RP_2, at the OC . Verify that that the system matrix, M, is given by

$$ M= \begin{pmatrix} 1 & T \\ 0 & 1 \end{pmatrix} \begin{pmatrix} 1 & 0 \\ -P_1 & 1 \end{pmatrix} \begin{pmatrix} 1 & T \\ 0 & 1 \end{pmatrix} \begin{pmatrix} 1 & 0 \\ -P_2 & 1 \end{pmatrix} $$

here $T= \dfrac{Lc-L}{nair} + \dfrac{L}{nrod}$

lve to get the A, B, C and D of the laser resonant cavity and then can lve for beam diverge, beam radius of curvature, as shown below. We are suming a fundamental Gaussian beam in the cavity.

2.3.3.1.1. Beam Divergence

The beam divergence can be calculated

$$\frac{1}{R} = \frac{D-A}{2B}$$

Where R is the radius of curvature of the wave front.

2.3.3.1.2. Beam Radius of Curvature

It also follows then that the radius of curvature of the beam is the reciprocal of the beam divergence, so R is given as

$$R = \frac{2B}{D-A}$$

2.3.3.1.3. Matrix Spur and Resonator Stability

Matrix Spur=A+D

If $1 > \frac{A+D}{2} > -1.$

then a resonator is stable, otherwise
it would be unstable

3.3.1.4. Propagation of a Gaussian Beam and its Complex Curvature

The complex curvature parameter can be written as

$1/q=1/R+i\lambda/\pi\omega^2$ (this is a complex expression)

Imaginary part($\lambda/\pi\omega^2$) is
a measure of the degree
of power concentration in
the axial region of the beam

Real part – divergence of surfaces of constant phase

Laser Advanced Concepts

2.3.4. Self-Test Ideal ABCD Matrix Systems

In an optical ABCD Matrix System what is the significance of

	Matrix Element A=0, Significance	True	False
1	Incident rays are collimated		
2	Exit rays converge at y_2		
3	Incident rays starting at height y_1		
4	Incident rays are collimated		
5	Exit rays are collimated		

	Matrix Element B=0, Significance	True	False
6	Incident rays are collimated		
7	Exit rays converge at y_2		
8	Incident rays starting at height y_1		
9	Incident rays are collimated		
10	Exit rays are collimated		

	Matrix Element C=0, Significance	True	False
11	Incident rays are collimated		
12	Exit rays converge at y_2		
13	Incident rays starting at height y_1		
14	Incident rays are collimated		
15	Exit rays are collimated		

	Matrix Element D=0, Significance	True	False
16	Incident rays are collimated		
17	Exit rays converge at y_2		
18	Incident rays starting at height y_1		
19	Incident rays are collimated		
20	Exit rays are collimated		

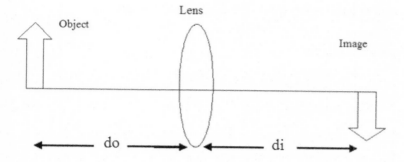

Figure T3.1. Object projector

Refer to the object-image system above for the next 4 questions

21. The Transfer Matrix between the lens and the image is

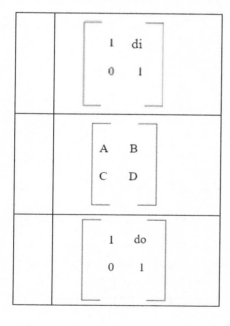

22. The Transfer Matrix between the object and the lens is

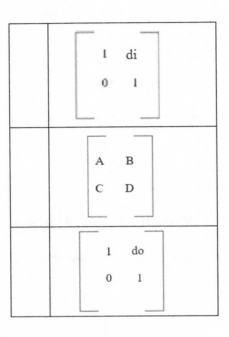

23. The system matrix M is

$$M= \begin{bmatrix} 1 & di \\ 0 & 1 \end{bmatrix} \begin{bmatrix} A & B \\ C & D \end{bmatrix} \begin{bmatrix} 1 & do \\ 0 & 1 \end{bmatrix}$$

The active medium of an optically pumped laser can act as a thermal lens, whose dioptic power (1/f) is proportional to the pump power. Below we depict the cavity configurations in such a cavity.

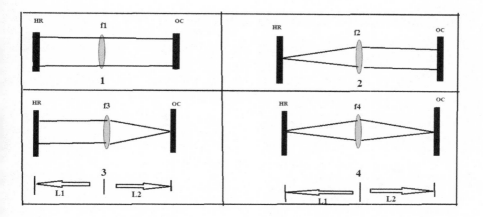

Figure T3.2. Active resonator representations

Assume the HR to be input and OC the output, and if you were to model these lasers resonate cavity configurations using ABCD matrices considering the thermal lensing of the active medium.

24. Configuration 1 would have matrix element
    a. A=0
    b. B=0
    c. C=0
    d. D=0
    e. None of the above

25. Configuration 2 would have matrix element
    a. A=0
    b. B=0
    c. C=0
    d. D=0
    e. None of the above

26. Configuration 3 would have matrix element
    a. A=0
    b. B=0
    c. C=0
    d. D=0
    e. None of the above

27. Configuration 4 would have matrix element
    a. A=0
    b. B=0
    c. C=0
    d. D=0
    e. None of the above

28. Given an optical system described by an ABCD matrix such that the element C is equal to zero (C=0) then the input (left) and output (right) rays are as shown in

Input Rays   Output Rays

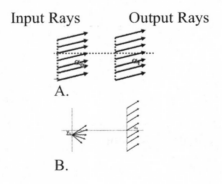

A.

B.

C. Any of the above
D. None of the above

29. In the optical system shown below

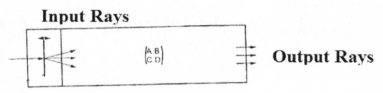

Input Rays

$\begin{pmatrix} A & B \\ C & D \end{pmatrix}$

Output Rays

Which matrix element is zero?

A. A

B. B.

C. C

D. D

E. None of the above

Laser Advanced Concepts

4. *Laser Gaussian Beams and Diffraction-Limited Focusing*

Topics

4.1. Laser Gaussian Beam Propagation and Characterization
4.2. Laser Beam Collimators/Expanders/Telescopes
4.3. Laser Gaussian Beam M^2
4.4. Diffraction-Limited Focusing of Gaussian Beams
4.6. Laser Gaussian Beams and Diffraction-Limited Focusing

4 .1. Laser Gaussian Beam Propagation and Characterization

Gaussian Beam Propagation and its Complex Curvature

- Gaussian beam is the term used to describe a diffraction-limited beam of coherent radiation, whose energy remains concentrated near the axis of propagation and falls off rapidly according to a Gaussian function.
- The radius ω (we usually call r_1) is the a "spot radius" at which the amplitude falls to $1/e$ and the irradiance to $1/e^2$ of the central value.

    Below we depict a focused laser beam where we specify some of the beam parameters.

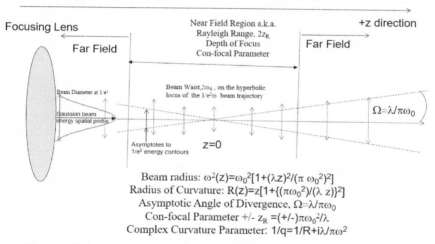

Beam radius: $\omega^2(z)=\omega_0^2[1+(\lambda z)^2/(\pi\,\omega_0^2)^2]$
Radius of Curvature: $R(z)=z[1+\{(\pi\omega_0^2)/(\lambda\,z)\}^2]$
Asymptotic Angle of Divergence, $\Omega=\lambda/\pi\omega_0$
Con-focal Parameter $+/-\ z_R =(+/-)\pi\omega_0^2/\lambda$
Complex Curvature Parameter: $1/q=1/R+i\lambda/\pi\omega^2$

Figure 4.1. Focusing of a Gaussian beam.

- As the Gaussian beam is propagated in space, the effects of diffraction cause it to expand and diverge slowly i.e. both
    - spot
    - curvature R vary slowly with the +z direction
- From the wave equation it can be shown that
- spot size, $w^2(z)=w_0^2[1+(\lambda z)^2/(\pi\omega_0^2)^2]$
- Radius of curvature, $R(z)=z[1+\{(\pi\omega_0^2)/(\lambda z)\}^2]$

- If a Gaussian beam is focused the curve representing the locus of $1/e^2$ is a hyperbola whose closest approach to the z-axis is ω_0 (radius) at $z=0$
- Whose asymptotes are at an angle
  - $\Omega=+/-\lambda/\pi w_0$ [Note that these are two half angles of divergence]
- The surfaces of constant phase are planes near the "neck" (beam waist) of the Gaussian beam and they acquire their strongest curvature at a distance
- $(+/-) z_0 = (+/-)\pi\omega_0^2/\lambda$ from the center.
- The central region of length $2z_0$ over which the cross-section remains nearly constant is referred to as the near-field.
- Beyond the near field is the far field in either direction
- Electric field intensity is highest at the beam waist and in the near-field and drops off as the beam width increases, especially in the far-field.
- Even though the field intensity varies, the total energy inside $1/e^2$ contours remains constant at 86%.
-

  - The complex curvature parameter can be written as
  - $1/q=1/R+i\lambda/\pi\omega^2$ (this is a complex expression)
    Imaginary part($\lambda/\pi\omega^2$) is a measure of, $1/\omega^2$, the degree of power concentration in the axial region of the beam
  - Real part – divergence of surfaces of constant phase

For Gaussian TEM$_{00}$ beams we shall let

      d = beam diameter at the beam waist

      ω_0 = beam radius at the beam waist

      D= beam diameter at a distance z

      ω= beam diameter at a distance z

      λ=wavelength

      z=distance from beam waist

- When focusing, a beam using a plano-convex lens the flat side faces the focused spot.

- **Beam Divergence**

    ○ If the beam width of an un-collimated Gaussian beam is transmitted through a positive lens, of focal length the angle of divergence, θ, of the beam is can be calculated as,

$\theta = \omega_f / f$

From Diffraction

Theory $\theta_{1/2} = \dfrac{\lambda}{\pi d_{1/2}}$

Beam Radius, ω_0, or Width

- The beam radius, ω, at a distance z is given size,
$\omega^2(z) = \omega_0^2[1+(\lambda z)^2/(\pi \omega_0^2)^2]$ where ω_0 is the beam radius at the beam waist and λ is the beam wavelength
- The energy intensity distribution in a TEM$_{00}$ resembles a Gaussian/Normal/Bell Curve in the Far-Field.

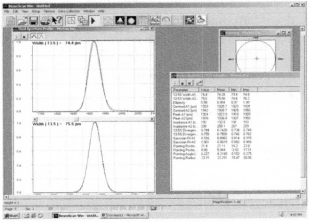

The shown Gaussian profiles are in the x-z and y-z planes . In measuring beam width or diameter we are only interested in transverse (TEM i.e. Transverse Electromagnetic] or cross-sectional sizes of the beam so we only measure the x and y widths. Usually the average of these two measurements in this x-y plane of the beam is submitted as the final width. This then makes the beam diameter a two dimensional profile as you mentioned.

$I = I_0 e^{-2(r/r_1)^2}$ *[Gaussian or Normal Curve Equation]

superscript for squared

When r=0 then $I = I_0$

When $r = r_1$ $I = I_0 e^{-2(1)^2} = I_0/e^2$

so

$I = I_0/(2.718)^2 = 13.5 \% \ I_0$

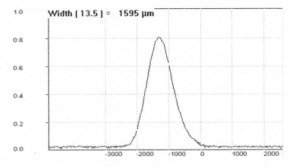

Width [13.5] = 1595 µm

- The total energy inside $1/e^2$ contours of a Gaussian beam remains constant at 86.5%.

- A Gaussian laser beam diameter/width is by default measured at 13.5 (=$1/e^2$)
- Given a Gaussian beam whose transverse profile exhibits a D_y diameter of 100 microns (i.e. 100 x 10^{-6} mm) and a D_x diameter of 102 microns. The beam ellipticity/roundedness/circularity is 1.02 or .98 depending on which diameter is the numerator or denominator, both numbers convey the fact that the beam's x-y profile is 2% from being a perfect circle.

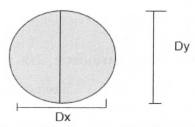

Figure 4.3. Depiction of the transverse profile of a laser beam

- **Beam Waist, (@z=0**
  - ○ When collimated light is passed through a convex lens it will converge at the beam waist.
  - ○ The wave-front is planer at the beam waist
  - ○ Beam waist is also called beam spot size
  - ○ A beam waist is only located at the focal point only if the incoming beam is collimated.

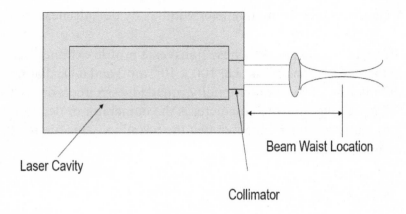

Beam Waist Location

Laser Cavity

Collimator

Figure 4.4. Laser beam collimator inclusion in a laser.

- **Depth of Focus/Con-focal Parameter, $2z_R$**

    o The Depth of Focus ($2Z_R$) is a region near the beam waist, over which the laser beam remains relatively collimated
    o A Gaussian beam acquires its strongest curvature at a distance
    o $(+/-)\pi\omega_0^2/\lambda$, at the edge if the Depth of Focus, from the beam waist
    o The beam diameter at the edge of the Depth of Focus, +/-) Z_R is given by $D_{ZR} = \dfrac{2w0\sqrt{2}}{1}$
    o Therefore $Z_R = \pi\omega_0^2/\lambda$

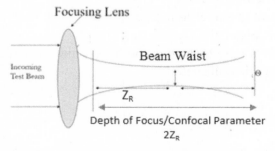

Focusing Lens

Beam Waist

Incoming
Test Beam

Z_R

Depth of Focus/Confocal Parameter
$2Z_R$

Figure 4.5. Illustration of a focused beams depth of focus.

    o Rayleigh Range is half the Depth of Focus i.e. Z_R

- **Far-Field**
  - The region beyond the near field, on either side, is called the Far-field
  - Far Field Diffraction is also called Fraunhofer Diffraction
  - Screen distance Z'>100 (Area of aperture opening/wavelength of incident light
  - Z' is usually too far way so we use a positive lens to create a far-field just to the right of the beam waist.
  - All laser beam tests and measurements are done in the far-field
  - A positive lens is used to "bring" the far-field within a lab workbench

Also see Near-Field

- **Half Angle of Divergence, Ω**
  - As the beam aperture size increases beam divergence decreases

- **Near-Field**
  - In the region centered at the beam waist of a focused Gaussian beam is referred to as Near-Field
  Near Field Diffraction is also called Fresnel Diffraction
  - Screen distance Z'<100 (Area of aperture opening/wavelength of incident light

Also see Far-Field

- **Radius of Curvature**
  A focused Gaussian beam has a spherical radius of curvature, R, given by

$$R(z) = z[\ 1 + (\pi\omega_0^2/\lambda z)^2]$$

Where z is the location where the beam is being evaluated

ωo is the beam radius, and

λ is the wavelength of the beam.

- When the radius of curvature of a beam goes to infinity the beam is collimated
- The radius of curvature is numerically equal to infinity at the beam waist (z=0).
- The wave-front has a positive radius of curvature before the beam waist
- The wave-front has a negative radius of curvature after the beam waist

Propagation Formulae

	Gaussian/TEM$_{00}$ Beam
Beam Radius	$\omega^2(z)=\omega_0^2[1+(\lambda z)^2/(\pi\omega_0^2)^2]$
Beam Waist (Radius at z=0)	ω_0
Depth of Focus/Con-focal Parameter	$2z_R$
Half Angle of Divergence, Ω	$\Omega=\lambda/\pi\omega_0$
Radius of Curvature, R	$R(z)=z[1+\{(\pi\omega_0^2)/(\lambda z)\}^2$
Rayleigh Range	$z_R=\pi\omega_0^2/\lambda$

Personal Notes

2.4.2. Laser Beam Collimators/Expanders/Telescopes

- Effects of diffraction cause a Gaussian laser beam to expand as it propagates in media.
- If a laser beam size is increased, then its angle of divergence will decrease, and this is what is usually done in a laser beam collimator/telescope
- Collimators are always inserted at the exit port of a laser

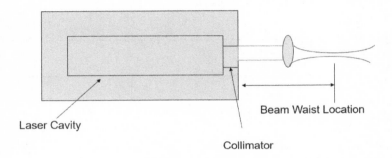

Figure 4.6. Imaging of a resonant cavity beam waits by a focusing lens.

The Principle of Reversibility and Collimators
- Collimators exploit the fact that rays emanating from the focal point of a positive lens will be collimated by the lens, i.e. the reverse/inverse of
"Rays from infinity/collimated rays are focused at the focal point of positive lens".

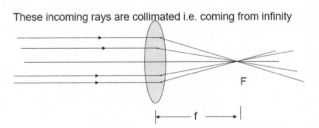

A collimated laser beam will converge after passing through a convex lens.
The refracted rays will cross each other at the focal point.
What happens if we reverse the rays?

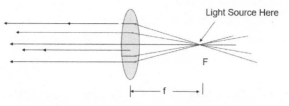

Light Source Here

F

|←—— f ——→|

What happens if we reverse the rays with the rays at the focal point?
Rays should follow the same trajectory on the way back.

Divergent light rays, such as from a laser diode/diode laser can be collimated by placing a positive lens exactly one focal length from it as shown below

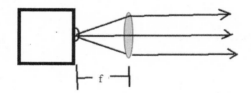

Figure 4.8.   Point-source beam collimation using The Principe of Reversibility.

- In practical environments the collimation can be optimized using autocollimation techniques as shown below.

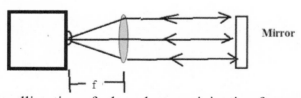

Figure 4.9. Autocollimation of a laser beam originating from a point source.

- If the rays going toward the mirror are collimated, upon reflection they will focus
- Some lasers' beams cannot be treated as point sources so two-lens collimators would be needed as we shall see below.

97

From diffraction theory

The angle of divergence is $\theta = k\lambda/d$

Where k is a constant and is $4/\pi$ for Gaussian beams

So

$\theta d = k\lambda$

The right side of this equation, $k\lambda$, is a constant so

If a beam is expanded, then its angle of divergence will decrease and vice versa.

These are some of the benefits of installing laser collimators in lasers.

- A Keplerian Collimator is composed of two positive lenses

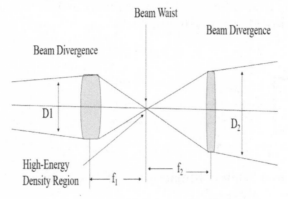

Figure 4.9. Cartoon of a Keplerian beam expander/collimator.

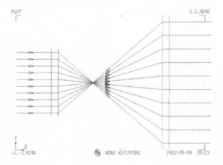

Figure 4.10. 2-D matrix simulation of the output of a Keplerian beam expander/collimator. [*Graphics produced using ASAP™ (Breault Research Organization.]*

- A Galilean Collimator is composed of one negative and one positive lens

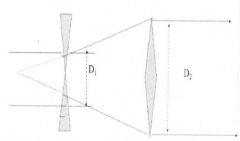

Figure 4.11. Cartoon of a Galilean beam expander/collimator.

- A Galilean Collimator is ideal for high energy beams since it does not increase the irradiance/fluence.
- High energy density laser beams could cause ionization of gases in air resulting attraction of dirt to within it.
- Collimators/Expanders or Telescopes are needed at the exit aperture of a laser in order to achieve an ideal laser beam with low divergence and M^2 value (<1.3)
- The two lenses in a collimator separated by a distance approximately equal to the sum of them to the sum of their focal lengths.
  - Approximate because the incoming beam from the laser cavity is not collimated, so it will not focus at the focal point in the case of a Keplerian Collimator, and it will not be diverged such that it appears to be emanating from the focal point of the negative lens in the case of a Galilean Collimator.
- The output beam size of a laser beam collimator/telescope D_2 is given as
$D_2 = D_1|f_2/f_1|$
where
D2 is the expanded beam
D_1 is the unexpanded beam diameter
  - The exit beam size, D_2, from a collimator is given by
    - $D_2 = D_1|f_2/f_1|$
  - Laser beam collimators are always used as beam expanders

99

as well because it makes the beam less divergent

- The input D_1 is expanded by the ratio $|f_2/f_1|$ (=X) hence D_2 is always greater by the ratio X

  Sometimes laser beam collimators are marketed at 2X, 3X, 4X...... NX telescopes

Personal Notes

2.4.2. Self-Test Laser Beam Collimators/Expanders

1. If a collimated Gaussian laser beam of wavelength λ is passed through a positive lens of focal length f then the beam waist will be located at a distance____ from the lens

   a. f

   b. λ

   c. θ

   d. None of the above

2. If a collimated Gaussian laser beam of width d has its width expanded, then its angle of divergence will
   a. increase
   b. decrease
   c. none of the above

3. Collimators/Expanders/Telescopes are needed at the exit aperture of a laser in order to achieve an ideal laser beam with low
   a. Divergence.
   b. M^2 value (<1.3)
   c. All the above
   d. None of the above

The next three questions are based the laser head characterized below. You are to assemble a laser head from the clean room into a laser system. As you receive the laser you note, among other things, that it has an M^2 value of 1.3.

4. As you align the laser system you find out that it is outputting the second and third harmonics of 1064 nm but the system only needs the third harmonic. Which dichroic mirror/reflector would you use to get rid of 1064 nm and the second harmonic?
   a. 1064 nm dichroic mirror/reflector
   b. 355 nm dichroic mirror/reflector
   c. 532 nm dichroic mirror/reflector
   d. 266 nm dichroic mirror/reflector
   e. None of the above

5. If after retaining the third harmonic (from the previous question) you realize that the beam is highly divergent, which optical component/system you would utilize to fight/reduce the divergence.
   a. microscope
   b. diffraction grating
   c. collimator
   d. convex lens
   e. None of the above

6. After the divergence reducing, you expect the M^2 value to be
   a. less than
   b. greater than
   c. equal to 1.3

7. The spot size d of a laser beam focused by a lens system of focal length f is given by
   a. $d=f\theta$
   b. $d=M\theta$
   c. $d=Mf$
   d. $d=1/f$
   e. None of the above

8. If a Gaussian laser beam of wavelength λ is passed through a positive lens of focal length f, and the beam width at f is d, then angle of divergence, θ, of the laser beam is given by
   a. $\theta=d/f$
   b. $\theta=\lambda/f$
   c. $\theta=f/d$
   d. $\theta=\lambda/f$
   e. None of the above

9. If the same beam is both collimated and then un-collimated and both beams are focused by a positive lens, will their beams waists lie at the same distance from the lens.
   a. True
   b. False

10. M^2 decreases as beam divergence, θ, decreases.
    a. True
    b. False

11. Figure below is a matrix simulation of a _____ collimator/expander

Graphics produced using ASAP™ (Breault Research Organization).

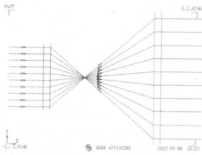

Figure T4.1.

   a. Galilean
   b. Keplerian
   c. M^2
   d. Gaussian
   e. None of the above

12. Figure below depicts the matrix simulation of a laser expander/collimator with a __output.

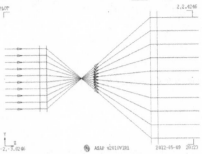

Figure T4.2.

a. 1X
b. 2X
c. 3X
d. 4X
e. None of the above

13. Figure below is a matrix simulation of a ____ collimator/expander.
((Graphics produced using ASAP™ (Breault Research Organization).

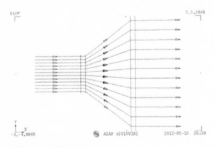

Figure T4.3.

a. Galilean
b. Keplerian
c. M^2
d. Gaussian
e. None of the above

14. Given a highly divergent point source and a positive lens of focal length 100 mm, what should the distance between the lens and the source, do, be for the output rays to be collimated?

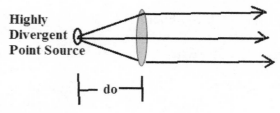

Figure T4.4.

a. 10 cm
b. 5 cm
c. Infinity
d. 20 cm
e. None of the above

15. If the rays going toward the mirror are collimated, upon reflection they will focus

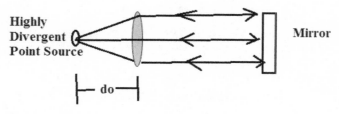

Figure T4.5.

a. to the right of point source
b. to the left of point source
c. at point source
d. we do not have enough information to know this

2.4.3. Laser Gaussian Beam M^2

- The M^2 value of a beam is also referred to as its
  - Beam Quality
  - Focusability Factor

M^2 embodies information on both beam divergence and waist

- $M\theta = \Theta$
- $Md = D$

θ is for a theoretical beam and smallest beam divergence achievable in reality and,

Θ is the divergence of a real beam.

d is for a theoretical beam and smallest beam waist achievable in reality and,

D is the diameter/width of a real beam.

M^2 represents a more rigorous way to determine how much a beam deviates from the fundamental/TEM_{00}

$M\theta = \Theta$

$Md = D$

$M^2 d\theta = D\Theta$ ⟹ $M^2 = D\Theta/d\theta$

Since $D => d$ and $\Theta => \theta$

$M^2 => 1$

$M^2 = 1$ is the best/lowest value possible

A truly Gaussian beam has an M^2 value of 1

- When computing M-squared, the following parameters are needed
  - Beam Diameter ($D = 2\omega_0$)
  - Beam Depth of Focus ($2Z_R$)
  - Laser wavelength λ

What M^2 Means

A single mode Gaussian TEM_{00} beam has an M^2 equal to 1

- A TEM_{00} ($M^2 =1$) beam when focused by a lens the **spot size (d) equals a minimum** defined by theory.
- All other other beams ($M^2>1$) will have spot sizes larger (D) than that of a $TEM_{00}/M^2=1$ (d)
- This M^2 value provided meaningful information especially if an applications involves small focused spot sizes
- Also, A $TEM_{00}/M^2=1$ has the **smallest possible beam divergence.**

- A M^2 value of 1 ($M^2=1$) means that a beam has the lowest angle of divergence and beam waist thus a diffraction-limited beam.
- M^2 is specified as a number equal, or greater than, one. A beam with an M^2 value of 1, $M^2=1$, is **perfect**

- **Importance/Meaning**
  - M^2 embodies/contains information about beams spot size and beam divergence.
  - A beam with $M^2=1$ has the both the smallest possible spot size and angle of divergence
    - For a given laser resonator, the smallest beam spot size is only possible if it is composed of the fundamental transverse or TEM_{00} mode
    - A truly TEM_{00} laser beam has a M^2 value of 1and Gaussian transverse intensity profile
  - The M^2 value of a laser beam can be used as an indicator of how well the beam will focus or diverge if an appropriate lens is put in its path
  - Some people refer to M^2 as the Focusability Factor
  - M^2 values greater than one are an indication of "contamination" of the fundamental mode by higher order modes

Personal Notes

2.4.3. Self-Test Laser Gaussian Beam M^2

1. A truly TEM$_{00}$ laser beam has a_____.
   a. M^2 value of 1
   b. Gaussian transverse profile
   c. a and b
   d. None of the above

2. A beam with a M^2 value of one (M^2=1) has the
   a. largest beam waist achievable
   b. largest angle of divergence achievable
   c. all the above
   d. None of the above

3. If a laser beam has a beam waist diameter of D$_0$ then the beam diameter/width at the either edge of the Rayleigh Range is given by
   a. $D_0\sqrt{2}$
   b. $2\sqrt{D_0}$
   c. 2D$_0$
   d. None of the above

4. A TEM$_{00}$ laser beam has the smallest possible beam _____.
   a. waist
   b. divergence.
   c. M^2 value
   d. All the above
   e. None of the above

5. The M^2 value of a laser beam can be used as an indicator of how tightly the beam will_____if an appropriate lens is put in its path.
   a. focus
   b. diverge
   c. All the above
   d. None of the above

2.4.4 Diffracted limited Focusing of Gaussian Beams

If a Gaussian laser beam of wavelength λ is passed through a positive lens of focal length f, and the beam width at f is d, then the spot size, d, is given by

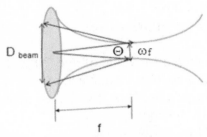

Figure 4.12. Dependence of beam size on the focal length of a lens.

- $\omega_f = f\Theta$ and it follows that
- $\Theta = \omega_f / f$

even if the beam is not collimated.

- For a given laser beam, the smallest beam spot size/waist is only possible if you have a TEM_{00} beam

A diffraction-limited Gaussian beam has a single (TEM_{00}) spatial mode

Diffraction Limited Beam Diameter

$M^2 = \omega\Theta/d\theta$

Recall from Diffraction Theory that $\theta = k\lambda/\omega$

$= \omega\Theta\pi/4\lambda$

but $\Theta = \omega_f/f$

So

$$M^2 = \frac{\omega\omega_f\pi}{4\lambda f}$$

$\omega = M^2 \, 4\lambda f/\omega_f\pi$

This is the smallest focused beam diameter for any Gaussian beam
If $M^2=1$ then it is

$$\omega= 4\lambda f/\omega_f\pi$$

This is the smallest possible

- In general, the beam waist size/spot size, ω, is directly proportional to the laser beam wavelength, λ, and focal length, **f,** of the focusing/positive lens
  - o Specifically, therefore the shorter the wavelength and focal length the shorter the spot size.

Diffraction-Limited Focusing, NA and f#

et's not consider the effect of the incoming beam size, D, on the spot size.

et's revere the far-field problem and let D_{beam} be a Gaussian spot size at e lens with focal length f that focuses a beam to a diameter ω_f

et $D_{beam}=D_{Lens}$

The numerical aperture, NA, of a lens is defined as

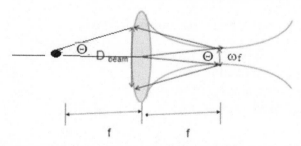

Figure 4.13. Illustration of Numerical Aperture variables in a focusing system.

$NA= \tan \Theta =(D_{Lens}/2f)$

$\Theta = \tan^{-1}(D_{Lens}/2f) = 4\lambda/\omega_f\pi$

113

i.e. Far-Filed angle of Divergence

Note that $\Theta = \tan\Theta = \sin\Theta$ for small angles.

$$D_{beam} = f\Theta = \text{(At Lens)}$$

Reacll that $\Theta = 4\lambda/\omega_f\pi$

So
$$D_{beam} = f4\lambda/\pi\omega_f$$

hence

$$D_{beam}\,\omega_f = f4\lambda/\pi$$

and the effective diameter of a focused beam would be

$$\omega_f = f4\lambda/D_{beam}\pi$$

The right side of this equation is a constant which means that expanding a beam before focusing will yield the tightest spot size

Recall that $NA = D_{beam}/2f$

so $\omega_f = 2\lambda/NA\pi$

Therefore, the lens with the largest numerical aperture, NA, will yield a beam with the tightest focus

The f-number of a focusing lens is defined as
$1/NA = f\# = 2f/D_{beam}$

- The $f^\#$ is also referred to as
    - Relative aperture OR
    - Speed of lens
- $\omega_f = f4\lambda/D_{beam}\pi$
- Let $D = D_{beam}\pi$ ... we will absorb π into the NA
  So

$$\boxed{\omega_f = 2\lambda(2f/D) = 2\lambda/NA = 2\lambda f\#}$$

Thus, an ideal Gaussian beam can be focused to within **two**

wavelengths times the f# of the focusing lens

We note from this preceding section that
- The lens with the largest numerical aperture, NA, will yield a beam with the tightest focus, ω_f.
- This is possible if a lens does not introduce aberrations. In practice a simple thin lens will not yield diffraction limited focusing unless used in combination with other lenses as doublets and triplets

Personal Notes

2.4.4. Self-Test Laser Gaussian Beams

and

Diffraction-Limited Focusing

1. The radius of curvature is equal to
   _____ at the beam waist.
   a. zero
   b. Infinity
   c. π
   d. a and c
   e. None of the above

2. The wave-front is _____ at the beam
   waist
   a. Curved
   b. Planar
   c. Any of the above
   d. None of the above

3. The wave-front has a _____ radius of
   before the beam waist
   a. Positive
   b. Negative
   c. Infinity
   d. None of the above

4. The wave-front has a _____ radius of
   curvature after the beam waist
   a. Positive
   b. Negative
   c. Infinity
   d. None of the above

5. When the radius of curvature of a beam
   goes to infinity the beam becomes
   _____.
   a. collimated
   b. convergent
   c. divergent
   d. None of the above

6. The beam radius, ω, at a distance z is given size, $\omega^2(z)=\omega_0^2[1+(\lambda z)^2/(\pi\omega_0^2)^2]$ where ω_0 is the beam radius at the beam waist and λ is the beam wavelength. What is the beam radius ω at z=0?
   a. ω_0^2
   b. ω_0
   c. $z\omega_0$
   d. None of the above

7. The curve representing the locus of focused Gaussian beam is a _____ whose closest approach to the z-axis is ω_0 (radius) at z=0
   a. parabola
   b. hyperbola
   c. a or b
   d. None of the above

8. A Gaussian beam acquires its strongest curvature at a distance _____ from the beam waist
   a. $(+/-)2\pi\omega_0^2/\lambda$
   b. $(+/-)\pi\omega_0^2/\lambda$
   c. $(+/-)3\pi\omega_0^2/\lambda$
   d. All the above
   e. None of the above

9. The region centered at the beam waist of a focused Gaussian beam is referred to as _____ -field.
   a. Near
   b. Far
   c. Middle
   d. All the above
   e. None of the above

119

10. The region on either side of the Near-Field, is called the _____ -field
    a. Far
    b. Mid
    c. Monochromatic
    d. All the above
    e. None of the above

11. The total energy inside $1/e^2$ contours of a Gaussian beam is _____%.
    a. 86.5
    b. 13.5
    c. Any of the above
    d. None of the above

12. If a Gaussian laser beam of wavelength λ is passed through a positive lens of focal length f, and the beam width a distance f

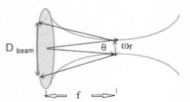

Figure T4.4.1

from lens is ω_f, then angle of divergence, θ, of the laser beam is given by
    a. $\theta = \omega_f/f$
    b. $\theta = \lambda/f$
    c. $\theta = f/\omega_f$
    d. $\theta = \lambda/f$
    e. None of the above

13. The beam size/width _____ distance from the laser
    a. increases with decreasing
    b. increases with increasing
    c. remains constant
    d. None of the above

14. As the beam aperture size increases beam divergence _____.
    a. increases
    b. remains the same
    c. decreases
    d. None of the above

15. A Gaussian laser beam diameter/width is by default measured at _____ of maximum laser intensity.
    a. 13.5
    b. $1/e^2$
    c. FWHM
    d. a and b
    e. None of the above

16. A laser beam with an M^2 value of one ($M^2=1$) has the smallest possible beam _____.
    a. waist
    b. divergence
    c. ellipticity
    d. a and b
    e. None of the above

17. The beam waist size of a focused laser beam depends on
    a. laser wavelength
    b. focal length of the focusing/positive lens
    c. a and b
    d. None of the above

18. Given a Gaussian beam whose transverse profile exhibits a D_y diameter of 100 microns (i.e. 100 x 10^{-6} mm) and a D_x diameter of 102 microns.

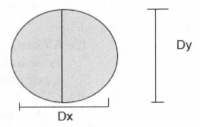

Figure T4.4.2. Depiction of the transverse profile of a laser beam

The beam ellipticity/roundedness/circularity is
    a. 1.02
    b. .98
    c. 1.0
    d. a and b
    e. none of the above

19. For a given laser resonator, the smallest beam spot size is only possible if you have a _____ beam.
    a. multimode
    b. higher order
    c. TEM_{00}
    d. Gaussian
    e. c and d
    f. All the above
    g. None of the above

20. Which one of these wavelengths can be focused by a positive lens to the smallest spot size?
    a. 532 nm
    b. 1066 nm
    c. 1064 nm
    d. 266 nm
    e. All the above
    f. None of the above

21. The far-field of a laser beam is too far out so a _____ lens is used to bring it within workbench.
    a. negative
    b. positive
    c. Any of the above
    d. None of the above

22. A laser beam's spot size/waist can be reduced by _____ the wavelength of a laser.
    a. increasing
    b. reducing
    c. any of the above
    d. None of the above

23. Fresnel diffraction is the _____ of a laser beam.

a. Near-field diffraction pattern
b. Far-field diffraction pattern
c. Grating diffraction pattern
d. Young's Double-slit pattern
e. All the above
f. None of the above

24. Fraunhofer diffraction is the _____.
a. Near-field diffraction pattern
b. Far-field diffraction pattern
c. Grating diffraction pattern
d. Young's Double-slit pattern
e. All the above
f. None of the above

25. Laser beam diagnostics are done in the_____ field.
a. near-
b. far-
c. a and b
d. None of the above

26. When collimated light is passed through a convex lens it will ae the _____ at the focal point of the lens.
a. smallest width
b. beam waist
c. a and b
d. None of the above

27. Beam waist is also called beam _____.
a. spot size
b. a and b
c. None of the above

28. Rayleigh range (Z_R) is a region near the beam waist, ω, over which the laser beam _____

    a. remains relatively collimated
    b. diverges maximally
    c. None of the above

29. The diameter at the end of the Rayleigh range, Z_R, is given by _____.
    a. $4\lambda/\pi$
    b. SQRT $(2)*\omega$
    c. All the above
    d. None of the above

    Where ω is the beam width/diameter at z=0.

30. TEM$_{00}$ beams exhibit a _____ transverse energy distribution.
    a. Bessel
    b. Voight
    c. Fresnel
    d. Gaussian
    e. None of the above

31. The region in which a focused Gaussian beam's cross-sectional area remains almost constant is known as the _____.
    a. Depth of Focus
    b. Con-focal Parameter
    c. Near field
    d. All the above
    e. None of the above

32. When computing M-squared, the following parameters are needed
    a. Beam waist (ω)
    b. Depth of Focus ($2Z_R$)
    c. Laser wavelength λ
    d. a and c only
    e. All the above
    f. None of the above

33. If the beam with of an un-collimated Gaussian beam is transmitted through a positive lens, of focal length f and produces a beam waist ω_f a distance from the from the lens then the angle of divergence, Θ, of the beam is can be calculated as
    a. $\Theta = \omega/\omega_f$
    b. $\Theta = 2f\omega$
    c. $\Theta = \omega/f$
    d. $\Theta = \omega_f/f$
    e. None of the above

34. The M^2 value of a beam is also referred to as its
    a. Beam Quality
    b. Focusability Factor
    c. Divergence
    d. a and b
    e. All the above
    f. None of the above

35. If a collimated Gaussian laser beam of wavelength λ is passed through a positive lens of focal length f then the beam waist will be located at a distance _____ from the lens
   a. f
   b. λ
   c. Θ
   d. None of the above

36. If a collimated Gaussian laser beam of width D_1 has its width expanded to D_2 ($D_2>D_1$) then its angle of divergence, Θ, will _____.
   a. increase
   b. decrease
   c. Any of the above
   d. none of the above

37. Collimators are needed at the exit aperture of a laser to achieve an ideal laser beam with low_____.
   a. divergence.
   b. M^2 value (<1.3)
   c. All the above
   d. None of the above

38. The spot size, ω, of a laser beam focused by a lens system of focal length f is given by
   a. $\omega=f\Theta$
   b. $\omega =M\Theta$
   c. $\omega =M^2\Theta$
   d. $\omega =1/f$
   e. None of the above

39. The far-field half angle beam divergence, $\Theta_{1/2}$, can be calculated as _____.

    a. $\theta_{1/2}=2\lambda/\pi$
    b. $\theta_{1/2}=\lambda/\pi\omega_0$
    c. $\theta_{1/2}=2\lambda/\pi\,\omega_0$
    d. All the above
    e. None of the above

Where ω_0 is the beam radius at the beam waist of diameter, ω.

40. The far-field full angle beam divergence, θ, can be calculated as
    a. $\Theta=2\lambda/\pi$
    b. $\Theta=\lambda/\pi\,\omega_0$
    c. $\Theta=4\lambda/\pi\,\omega_0$
    d. All the above
    e. None of the above

41. The numerical aperture, NA, of a lens is defined as _____ where

$$\theta = \tan^{-1}(D_{Lens}/2f)$$

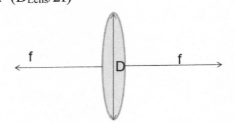

Figure T4.4.3.

    a. $NA= \sin\Theta$
    b. $NA= \tan\Theta$
    c. a or b
    d. None of the above

42. If a collimated and an un-collimated beam are both focused by a positive lens their beam waists will be at the _____ distance(s) from the lens.
    a. same
    b. different
    c. (this cannot be predicted)

43. Effects of diffraction cause a Gaussian laser beam to _____ as it propagates in media.
    a. Expand
    b. Shrink
    c. Collimate
    d. a and b
    e. All the above
    f. None of the above

44. The exit beam size, D_2, from a collimator is,
    a. $D_2 = D_1 |f_2/f_1|$
    b. $D_2 = f_1 |D_2/f_1|$
    c. $D_2 = f_1 |f_2/D_1|$
    d. all the above
    e. None of the above

45. A Keplerian Collimator is composed of _____ lens(es)
    a. two negative
    b. two positive
    c. one negative and one positive
    d. None of the above

46. A Galilean Collimator is composed of _____ lens(es)
    a. two negative
    b. two positive
    c. one negative and one positive
    d. None of the above

129

47. A _____ Collimator is ideal for high energy beams
    a. Keplerian
    b. Galilean
    c. a and b
    d. None of the above

48. High energy density laser beams could cause _____ of air inside a collimator.
    a. ionization
    b. heating
    c. a and b
    d. None of the above

49. Laser Beam Collimators/Expanders/Telescopes are needed at the exit aperture of a laser in order to achieve an ideal laser beam with low_____.
    a. divergence.
    b. M^2 value (<1.3)
    c. All the above
    d. None of the above

50. Figure below is a matrix simulation of a
_____ collimator/expander

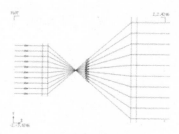

Figure T4.4.4.
(Graphics produced using ASAP™ (Breault Research Organization).

       a. Galilean
       b. Keplerian
       c. M^2
       d. Gaussian
       e. None of the above

51. Figure below depicts the matrix simulatio
of a laser expander/collimator with a
_____ output. (Approximate within +
1 mm)

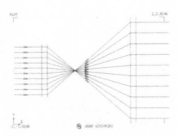

Figure T4.4.5.
(Graphics produced using ASAP™ (Breault Research
Organization).

a. 1X
b. 2X
c. 3X
d. 4X
e. None of the above

52. The two lenses in a collimator separate by
a distance approximately equal to the sum
of their _____.
   a. focal lengths
   b. radii of curvature
   c. a and b
   d. None of the above

53. Collimators can the diameter of a laser
beam
   a. expand
   b. decrease
   c. a and b
   d. None of the above

54. A collimator is ideal for high-power laser beams
    a. Galilean
    b. Keplerian
    c. Any of the above
    d. None of the above

55. Given a Keplerian Collimator constructed using two plano-convex lenses, which side of the first plano-convex lens should the incoming beam encounter first
    a. Plane
    b. Convex
    c. Any of the above

56. Given a Keplerian Collimator constructed using two plano-convex lenses, which side of the second plano-convex lens should the expanding beam encounter first
    a. plane
    b. convex
    c. Any of the above

57. Given a highly divergent point source and a positive lens of focal length 100 mm, what should the distance between the lens and the source, do, be for the output rays to be collimated?

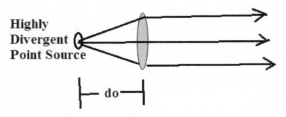

Figure T4.4.6.

    a.   10 cm
    b.   5 cm
    c.   Infinity
    d.   20 cm
    e.   None of the above

58. If the rays going toward the mirror are collimated, upon reflection they will focus

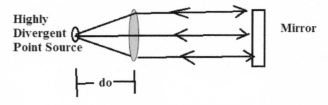

Figure T4.4.7.

    a.   to the right of point source
    b.   to the left of point source
    c.   at point source
    d.   we do not have enough information to know this.

FREE Laser IQ Scan/Test at laserpronet.com.
Click on Laser IQ Scan

Test - FREE Laser Tech 1 IQ Scan

▼ Test -

You are taking the following test:

Name   Test 0.1 FREE Laser IQ Scan1

Timer   Time Remaining (minutes:seconds) :

Your score, percentile etc. will be e-mailed
to you/your manager instantly/as soon as
you complete the test.

**Print and frame your e-mailed certificate and show it to your
manager/team lead to validate your laser knowledge and
competency.**

Made in the USA
Las Vegas, NV
23 February 2022

44471947R00080